4th International Conference on Accelerating Biopharmaceutical Development 2015

Scottsdale, Arizona, USA
1 – 4 March 2015

ISBN: 978-1-5108-1578-0

Printed from e-media with permission by:

Curran Associates, Inc.
57 Morehouse Lane
Red Hook, NY 12571

proceedings.com

Some format issues inherent in the e-media version may also appear in this print version.

Printed by Curran Associates, Inc. (2015)

AIChE
120 Wall Street, FL 23
New York, NY 10005-4020

Phone: (800) 242-4363
Fax: (203) 775-5177

www.aiche.org

Additional copies of this publication are available from:

Curran Associates, Inc.
57 Morehouse Lane
Red Hook, NY 12571 USA
Phone: 845-758-0400
Fax: 845-758-2634
Email: curran@proceedings.com
Web: www.proceedings.com

TABLE OF CONTENTS

INTEGRATION

Increasing need for responsiveness and flexibility in clinical and commercial bio-pharmaceutical supply chains

Prashant Yadav

William Davidson Institute
AT THE UNIVERSITY OF MICHIGAN

MICHIGAN
ROSS SCHOOL OF BUSINESS

1

Time to Market has been the Main Metric

Cost Efficiency Follows

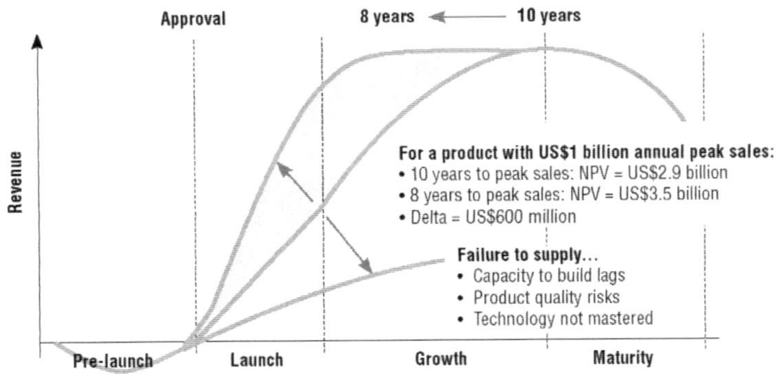

Source: IBM Business Consulting Services

1 day = For a \$500 million drug ~ \$1.5 million+
For a \$1 billion drug~\$3 million+

- **Cost of development is highly dependent on time to launch.**

Time to Market and FDA Breakthrough Therapy Designation

Breakthrough Therapy Designation Requests

	Requests Received	Requests Granted	Requests Denied
CDER (as of 6/14/2013)	58	21	16
CBER (as of 5/31/2013)	6	0	5
FDA Total	64	21	21

Source: FDA

Demand for affordability and improved access

- Intensified downward cost pressure as healthcare systems around the world struggle to balance increasing needs.
- The $100,000 product may not be sustainable –except may be the most innovative drugs
- To be competitive biopharma companies must revisit the pre-clinical, clinical and commercial manufacturing and supply strategies, technologies, and operational approaches they currently use
- Intrinsically in our decisions we have always faced the tradeoff
  - Reducing costs vs. Reducing time to market
  - We oscillate (within each firm and as a industry) on this tradeoff curve
- Is this necessarily a tradeoff?

4

Traditional model

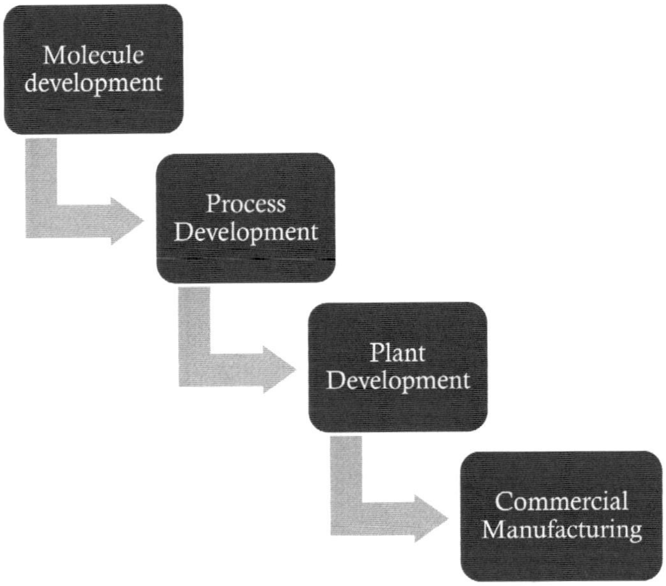

5

Assumptions in the early days of biopharma

- Early stage clinical manufacturing has different needs than commercial manufacturing
- Cost and throughput are not as important
- Flexibility and responsiveness are most important

6

Responsiveness, Agility and Cost Efficiency

Responsiveness: the ability to react quickly to satisfy unanticipated individual customer requirements

Agility: the ability to adapt capacities and processes to a change in the cost or market structure. Rapid reconfiguration ability

Cost Efficiency: the ability to reduce costs to the extent possible

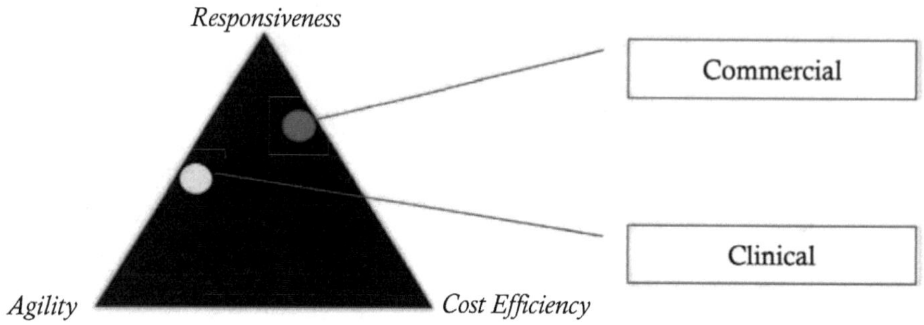

The clinical-commercial boundary starts to disappear, especially in biopharma

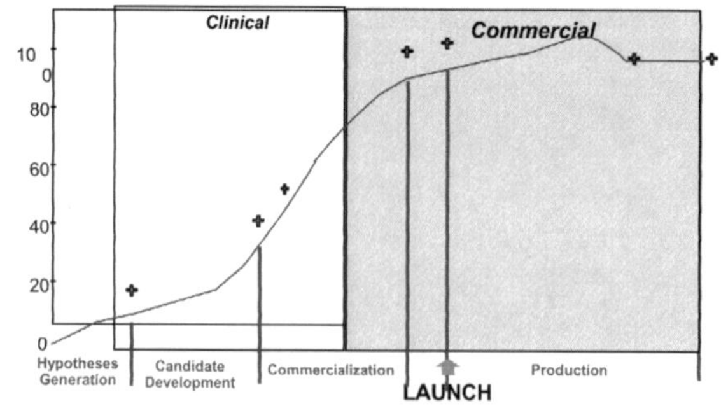

New generation technologies can support late stage process development, preclinical, clinical and limited commercial manufacturing

The change from "bow-tie" to "diamond"

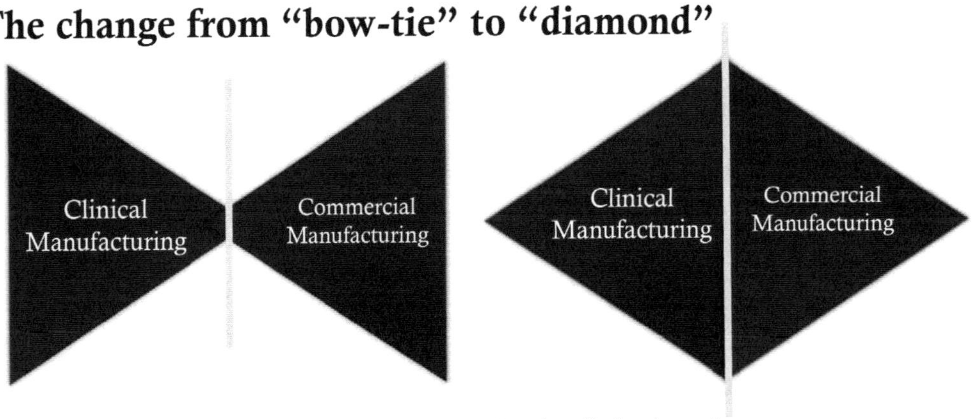

Create common key processes in both clinical and commercial supply chain

1. Global Capacity Balancing (including upstream and downstream capacity matching)
2. Demand Management
3. Inventory, Risk & Customer Service Level
4. Supply chain planning

9

Where does the new more merged structure operate?

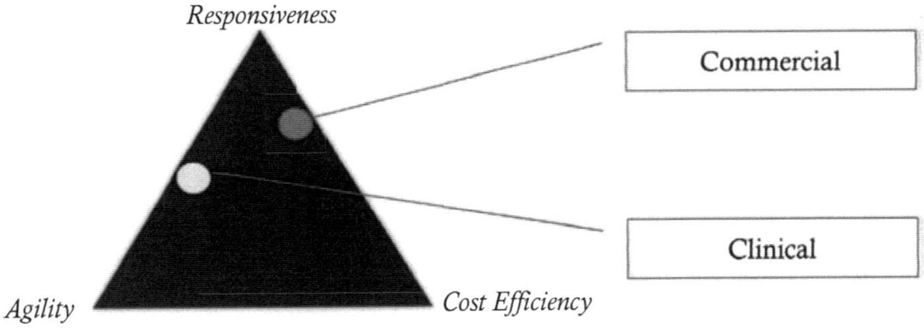

Is the (R,A,C) triangle still relevant?

- Flexible, modular, molecule agnostic manufacturing platforms may fundamentally change some of these tradeoffs.
- The new flexible manufacturing platforms may create significant shifts in the efficient frontiers

Challenges Faced by Clinical Supply Managers

Clinical Manufacturing and Supply Management became more complex

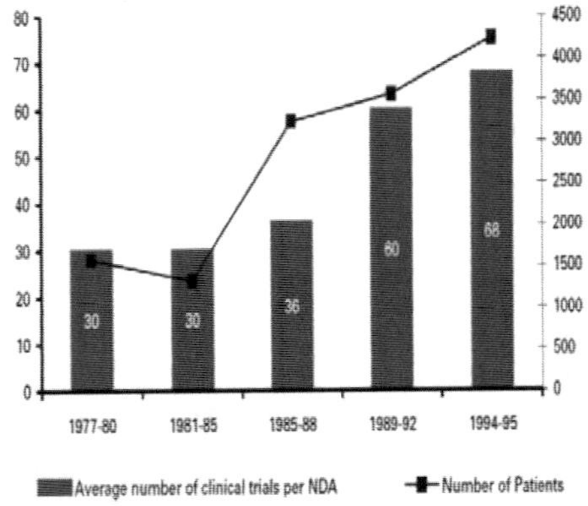

- More number of trials per NDA
- More patients per trial-patients tripled
- More complex material and protocol per trial
- Globalization of trials

Clinical Supply Management: Common Responses

- Increase efficiency-Decrease operating costs
- Integration across different parts of CSM
- Leverage technology for faster response
- Outsourced select functions of CSM
- Improve forecasting

In worrying about costs we worried about wastage

Bulk dose overage

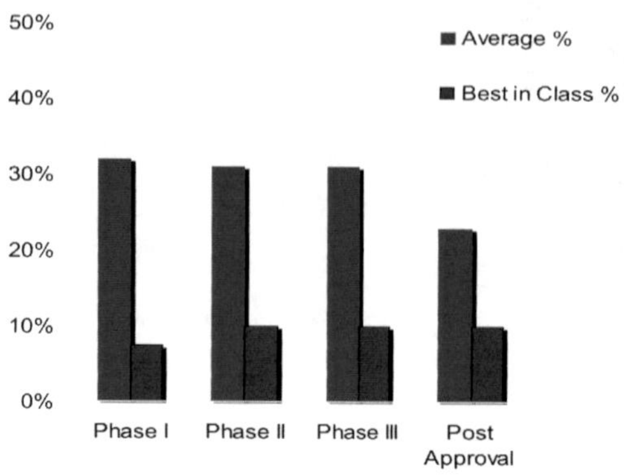

Patient kit overage

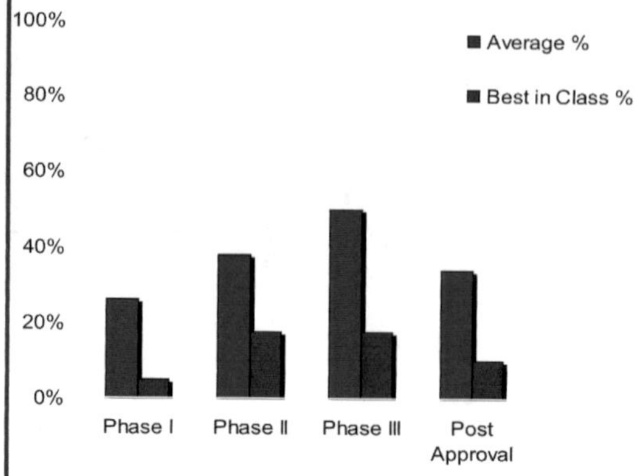

- Best in Class companies had lower overage

Source: Managing the Impact of Personalized Drugs on Clinical Trial Supply Operations, F nk Schulze, 2004

15

Regulatory pathways became different

- Slowly moving away from the binary approval process in favor of an incremental, *conditional* one.
- Drugs could be provisionally approved after promising early-stage data, with the FDA retaining the option to revoke that approval later on, should unexpected data come to light.

16

More Adaptive Trials & More Stratification = More challenges

- Considerable changes to the study as it advances
- Last minute decisions on dosage levels
- Resupply at variable intervals requiring variable safety stock at each site
- Need to coordinate delivery of different types of supplies to multiple active sites
- Inventory plans are developed one product at a time. Site operational plans are built for many products. Reconciling the two perspectives will become even more challenging

17

Breakthrough therapies: Additional challenges

- Breakthrough designation changes patient recruitment rates overnight
- Forecasting becomes more complex with such large swings in enrollment rates and drop percentages
- Higher risks of stockout
- Different expectations about CTM wastage
- Different safety stock assumptions

- This may creates complexity in planning, communication and especially decision making- slowing down
- Ability to reach key decision points quickly is the main driver for faster time to market.
- Defeats the purpose

18

Enhanced visibility across CSM

•Enhanced visibility throughout the network will become a pre-requisite to operate in this new environment
•Enhanced visibility will require standardized data across all clinical trial stakeholders.
•Visibility can't simply be improved via a tool.

Common data standards and improve visibility through tools and work practice modification

Demand and Supply Planning

Chemical development	Pharmaceutical development	Randomization, Packaging and Labeling	Stock positioning	Order Management	Patient allocation	Courier Flow Management

Role of technology

- Complex Studies will require more IT but a black box approach will not suffice.
- Deep understanding and "mastery" of systems, algorithms, simulation

Inputs
- Study protocol
- Sites
- Recruitment rate
- Dropout rate

Does not exist!!

Outputs
- Forecast demand for
  - IMP
  - Dosage form
  - Component materials
  - Drug substance
- Production schedule
- Distribution schedule
- Inventory Re order points
- Courier Shipping Plan

20

Is the (R,A,C) triangle still relevant?

- Flexible, modular, molecule agnostic manufacturing platforms may fundamentally change some of these tradeoffs.
- The new flexible manufacturing platforms may create significant shifts in the efficient frontiers

21

Trends and Technology that are shifting the tradeoff curves

- Low cost Low Volume
- Flexible /modular manufacturing
- Continuous bioprocessing
- Single use bio-reactors
- Process Intensification

22

Outsourcing and raw material supply

- Biopharma flexibility also involves make-or-buy decisions
- Adding biopharma contract-manufacturers can allow focusing on core technical skills and acquire/invest in assets appropriately
- Assurance of supply, change management, and business continuity planning become crucial
- Health of the raw material supply market including for new technologies e.g. single-use bioreactors is often an afterthought

23

Some ideas for consideration

- Use platform approaches wherever available and possible
  - Don't let Capital Efficiency/NPV drive you away from this
- Conduct Manufacturability Assessment as early in the development process
  - This needs to be communicated to development scientists
- High throughput process development from early stages
- Laboratory scale should mimic of large scale as far as possible
  - Technology-
  - Simulation
  - Statistics

24

More ideas for consideration

- Fast response and flexibility is not a fad
  - the ability to run multiple products in fewer lines,
  - the ability to respond quickly to the needs of a volatile market

will always be a significant source of competitive advantage

- Success will still be driven by
  - Process - What gets done Where
  - Performance Measurement- How is each sub unit performing
  - Organization - Who is responsible
  - Information Systems - Get information *in time* to make decisions

Achieving greater productivity in volatile demand environment requires improved fundamental scientific understanding of the manufacturing processes

Prashant Yadav
yadavp@umich.edu

International Conference on Accelerating Biopharmaceutical Development

Will Change in the Pharmaceutical Industry Change the Industry's Pecking Order?

Inno*Think*

Bernard Munos
Founder

Inno*Think* Center For Research in Biomedical Innovation
Scottsdale, AZ, March 1, 2015

1

The modern biopharma industry is 33 years old!

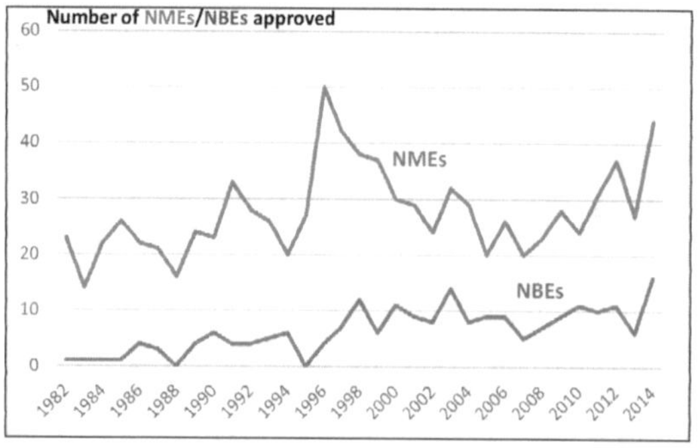

- **212 NBEs since 1982**
  - ➢ **23% of total (916)**
- **But,**
  - ➢ **33% of total in last 5 years**
  - ➢ **33% of total in last 10 years**
  - ➢ **34% of total in last 15 years**

Inno*Think*

2

Where is it going?

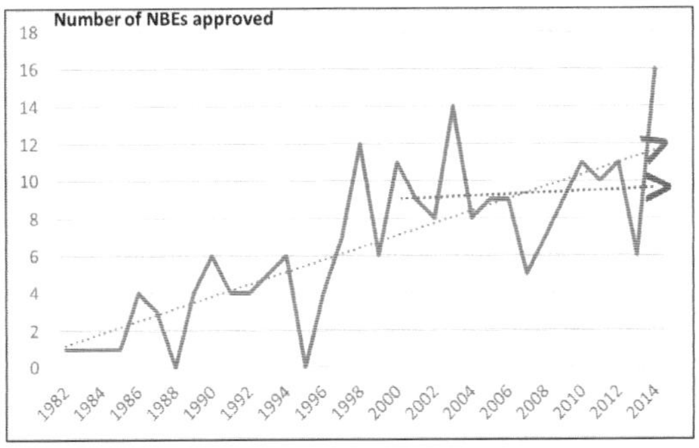

- **Are we stuck?**
- **Why?**
- **What to do?**

InnoThink

There is no shortage of opportunities

- **1982-2014**
  - Peptides
  - Monoclonals
  - recombinant vaccines
  - Antisense, fusion protein
- **2015-2025: same plus**
  - CAR T-cells
  - Conjugated antibodies
  - RNA therapeutics
  - RNAi
  - Stem cell therapies
  - Gene editing and other CRISPR applications
  - Syn bio
  - Oligonucleotides
  - Tissue engineering
  - Bioprinting
  - ...

The new face of biopharma

InnoThink

But there is also the reality

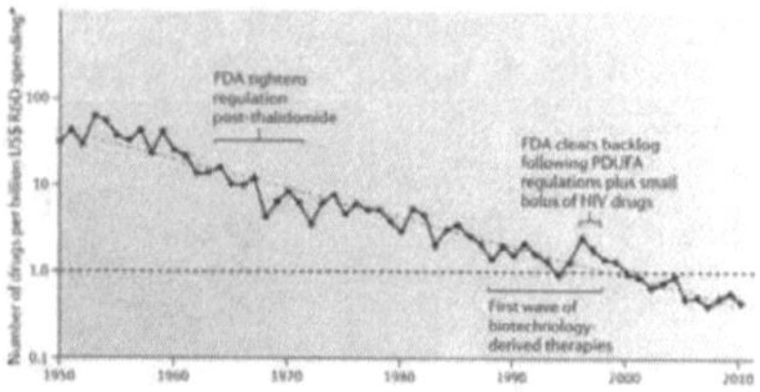

Eroom's law

SCIENTIFIC AMERICAN™

Cost to Develop New Pharmaceutical Drug Now Exceeds $2.5B

$2.6 billion

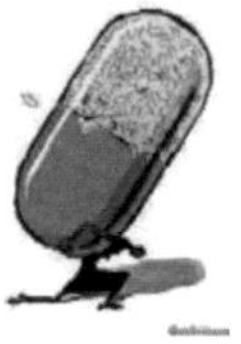

The $100,000 pill

Can biopharmaceuticals be made affordable?

InnoThink

There is also the disruptive potential

- **CAR T-cells**
- **Conjugated antibodies**
- **RNA therapeutics**
- **RNAi**
- **Stem cell therapies**
- **Gene editing and other CRISPR applications**
- **Syn bio**
- **Oligonucleotides**
- **Tissue engineering**
- **Bioprinting**
- ...

There is a lot of disruption here
- Disruption of research
- Disruption of manufacturing
- Disruption of marketing
- Disruption of medicine

Small companies love disruption
- Makes asymmetric warfare possible

Large companies generally do not
- Although some are warming up

Yet, it is disruption that made pharma great

InnoThink

But disruption is not enough to win

- **Disruption creates enormous value**
  - Genentech ($100 bn), Amgen ($120bn)
  - Microsoft, Google, Amazon, etc
- **Value can be captured by a few companies, or by the ecosystem**
  - Apple, Amazon, etc
  - Digital photography, DARPA (internet, GPS, e-publishing)

To win, you need disrupters + builders

InnoThink

There are two types of innovators

The "process" innovator	The "disruptive" innovator
Focused on growing and nurturing existing markets	Wants to obliterate what's there and replace it with something better
Seeks competitive advantage from greater efficiency	Seeks competitive advantage from changing the game
Improves and optimizes	Disrupts
Aligned with the organization's goals	Orthogonal to the organization's goals
Analytical, cautious	Intuitive, bold
Sticks to job description	Knows no boundaries
fits in	sticks out
Risk-averse, change-wary	Risk-taking, change-friendly
Seeks consensus	Never minds being alone

Source: adapted from Munos, Clin Pharmacol Ther. 2010 May;87(5):534-6

Process and disruptive innovators are different species; both are needed!

The crazies, the misfits, the rebels, the troublemakers, the square pegs...

InnoThink

8

You need both kinds of innovators to build great companies

"process" innovators	"disruptive" innovators

Big Pharma
Kodak
IBM

Money machines that will run out of steam

Many biotechs and device companies
Israel

Amazing technologies that won't realize their potential

Genentech
Amazon
Apple

You need both kinds of innovators to build great companies

The problem is
that they hate each other

So, what to do?

If you are a big pharma...

You need to welcome back disruptive innovators into your fold

- and create a climate where they can thrive
- protect them, celebrate them, and offer them career tracks

Or you need to reach out to them somehow

Various models

- The fiercely independent R&D division (Genentech, Novartis)
- The outreach organization (J-Labs, Pfizer's CTIs)
- The biotech-like ecosystem (GSK's DPUs)
- The collaborative pharma

InnoThink

If you are a small pharma...

You must put builders on your team...

- and eventually let them run the company
- while creating safe havens for your disruptive innovators

Various models

- Hire experienced pharma execs (who get it)
- Staff your board with like-minded pharma veterans
- Search for the rare bird (disruptive innovator + builder)

InnoThink

So what will happen?

- The odds are that pharma will be deeply disrupted
  - Industries can put off change, but they cannot avoid disruption
  - Biopharma will be a big driver
  - But other technologies (digital health), new business models (drug repurposing, crowdsourcing), and payer/patient pressure will lend a hand
- Big pharma that have reconnected with disruptive innovation will remain in the game if they can tackle the affordability challenge
- A new crop of small pharma will rise to exploit the opportunities offered by biologicals and other emerging technologies
- The new therapies will be affordable, better, and they will dwarf in number and value what the industry now brings

How fast it happens, and how much value it creates hinges upon you

InnoThink

Thank you!

Questions?

(bhmunos@stanfordalumni.org)

Biologics Beyond Today's Markets: The Case for Innovation in Bioprocess R&D to Enable Global Access of Biologics

Hadley, S.,

Originally presented on: 3/1/2015 0:50:00 - 1:15:00

Stephen Hadley is a Senior Program Officer for Vaccine Development (CMC) in Global Health at the Bill & Melinda Gates Foundation. In this role he provides technical CMC support to foundation partners involved in product development and makes investments in innovations in bioprocess R&D. The majority of projects supported by Steve are product or technology investments in recombinant proteins and monoclonal antibodies. Before arriving at the foundation, he was VP of quality for CMC ICOS Biologics, a biologics contract development and manufacturing organization, and was responsible for oversight of the QA, QC and analytical/formulation development activities for the company. Prior to CMC, Steve worked for ICOS Corporation as VP of quality and small molecule manufacturing (API and dosage form manufacturing). Steve worked as a research scientist with the FDA, where he was responsible for evaluating analytical methods for regulatory actions and for performing preapproval inspections of biotechnology companies. He also held scientist positions in discovery research and analytical quality control at NeoRx Corporation. Steve received his Ph.D. in Organic Chemistry from the University of Washington and BS degree in Chemistry from the University of Maine.

Amgen's Next Generation Biomanufacturing Facility

Hall, K.,

Originally presented on: 3/2/2015 0:00:00 - 0:25:00

Over past decades, the biopharma industry's emphasis has been on optimizing traditional cell culture platforms using stainless steel bioreactors to manufacture proteins. At Amgen, we've taken a different approach by investing in Amgen's Next Generation Biomanufacturing. The facility at Amgen Singapore is our first investment in next-generation flexible manufacturing. I will illustrate the facility's new design concepts and discuss the speed at which the facility was constructed. The result is a highly reconfigurable manufacturing facility that requires little or no retrofit, beyond specific new unit operations, to accommodate new product introductions.

Integrated Strategy to Enable Rapid Delivery of Material for IND Enabling Toxicology Studies

Estes, S., Biogen Idec
Ryll, T., Biogen Idec
Originally presented on: 3/2/2015 0:25:00 - 0:50:00

In an increasingly competitive business landscape, accelerating the ability to reach key decision points while minimizing the resources needed to do so has become a necessity for organizations to succeed. This has been evident in the biopharma industry with the major push over the years to achieve clinical proof of concept as rapidly as possible. Development of a manufacturing cell line and the production of representative material for toxicology is a key initial milestone in early clinical development and historically has represented a significant investment in time; frequently in excess of a year. In an effort to accelerate our IND readiness, we have reevaluated our approach to generating cell lines and producing Tox material. Through a holistic approach in which we have integrated strategic, operational and technical changes to our early clinical development work flow, we have created a new path in which it is possible to deliver representative material for IND enabling safety studies 7 months after selecting the final sequence for a candidate therapeutic. This has been enabled by the paradigm shift in which the material for Tox studies is generated prior to selection of the final cell line. To segue this approach from vision to reality, it was necessary to invest in improved expression systems and analytical capabilities necessary to rapidly and reliably produce sufficient quantities of well characterized drug substance.

REGENERON
science to medicine®

Speed to Clinic Powered by the VelociSuite® of Technologies

Darya Burakov, PhD
Associate Director, Protein Expression Sciences

4th International Conference on Accelerating Biopharmaceutical Development

---

REGENERON

2012, 2013 and 2014 Science Magazine: World's #1 Biopharmaceutical Employer

Corporate Headquarters- Tarrytown, NY

Industrial Operations and Product Supply
Rensselaer, NY

Satellite Office Bridgewater, NJ

REGENERON
science to medicine®

Regeneron Product Development Pipeline

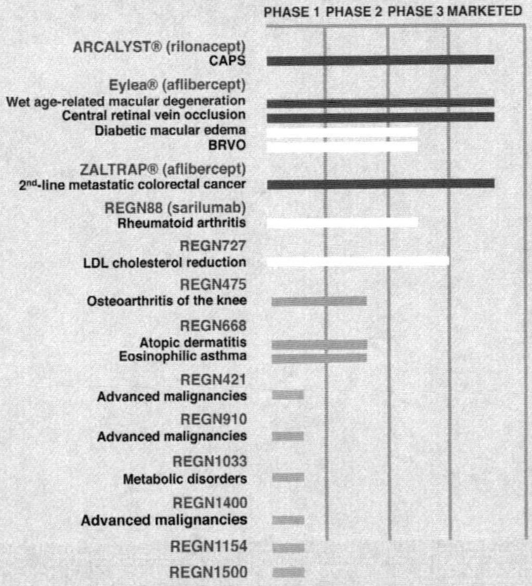

- ARCALYST® (rilonacept)
  - Approved for treatment of CAPS

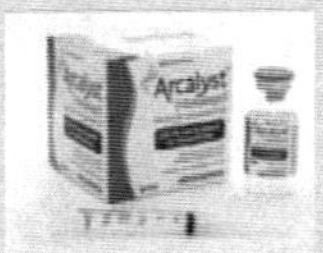

- Eylea® (with Bayer)
  - Approved for Wet AMD
  - Approved for CRVO

- ZALTRAP® (with Sanofi)
  - Approved for CRC
- 10 Fully-Human Antibodies in Clinical Trials (7 with Sanofi partnership)

The Global Spread of Infectious Diseases Calls for a Rapid Response

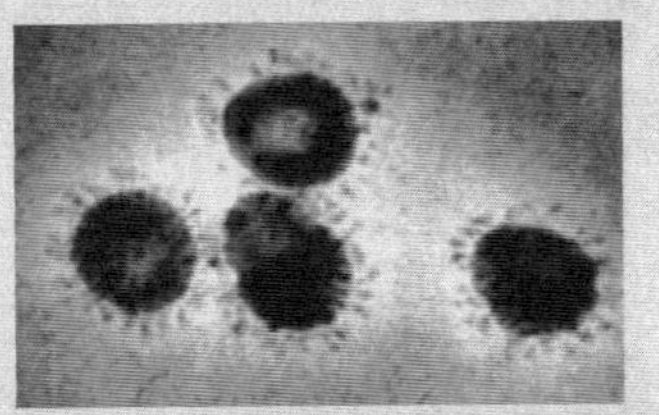

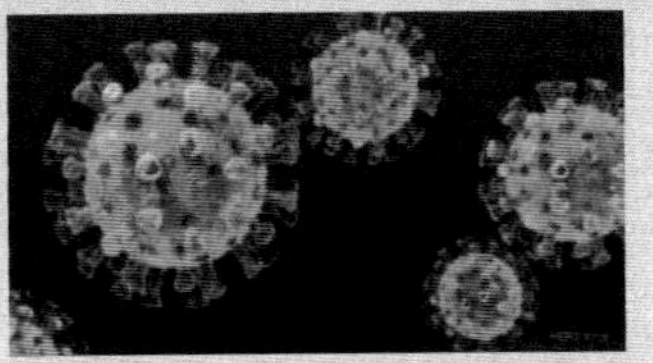

SARS: Severe Acute Respiratory Syndrome coronavirus; 2002-2003 outbreak in Asia caused 8,096 cases and 774 deaths reported in multiple countries. Within weeks, SARS spread from Hong Kong to infect individuals in 37 countries in early 2003.

MERS: Middle East Respiratory Syndrome coronavirus was first identified in Saudi Arabia in 2012; outbreaks have been reported in 22 countries, with over 600 cases and over 180 deaths.

Ebola: 2014 outbreak in West Africa is the largest in history, affecting multiple countries, 21,206 reported cases resulting in 8,386 deaths.

Infectious diseases spread rapidly in 2015

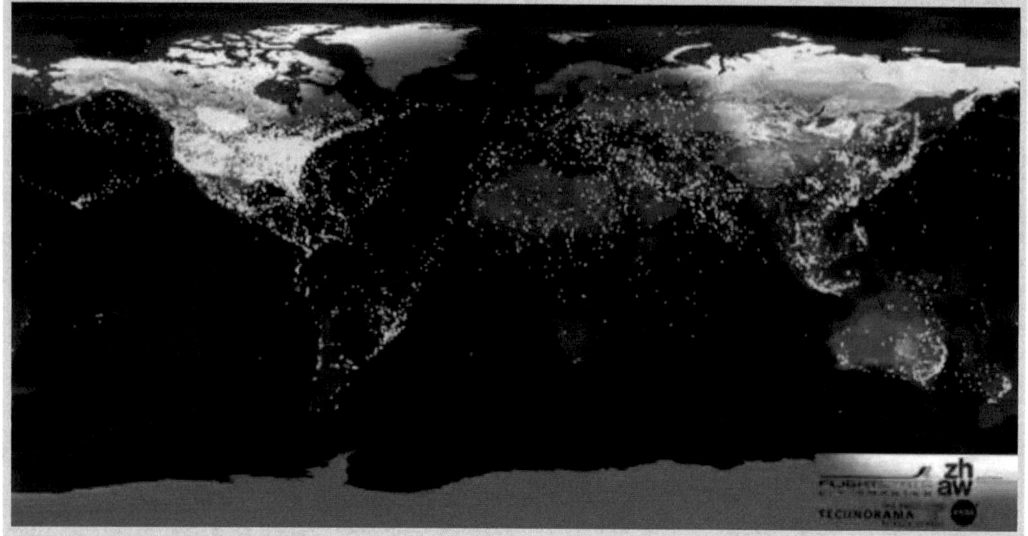

REGENERON
science to medicine

Rapid Response to Emerging Infectious Diseases

- **The existing therapeutic antibody development paradigms are incapable of rapidly responding to emerging infectious diseases**
  - Identification of therapeutic antibody candidates
    - Single antibody vs. antibody "cocktails"
  - Screening for efficacy in relevant models
  - Cell line development
  - Process development and production
- **Significant time reductions required for all steps in development process**

Regeneron's VelociSuite of technologies address these weaknesses

REGENERON
science to medicine

Regeneron's Technologies are the Foundation for the Company's Drug Discovery and Development

- **The VelociSuite® of proprietary technologies is the engine for development of fully-human therapeutic antibodies**
- **VelociGene®**
  - High-throughput generation of any desired genetic alteration in mouse embryonic stem cells
- **VelociMouse®**
  - High-throughput generation of mouse models directly from embryonic stem cells
- **VelocImmune®**
  - Mice mount robust immune response expressing human antibodies
- **VelociMab®**
  - EESYR®, FASTR™, NICE™, and BST for unparalleled high throughput speed from mouse to clinic

REGENERON
science to medicine

Regeneron *VelocImmune®* Mouse Rapid Generation of High Affinity Antibody Candidates

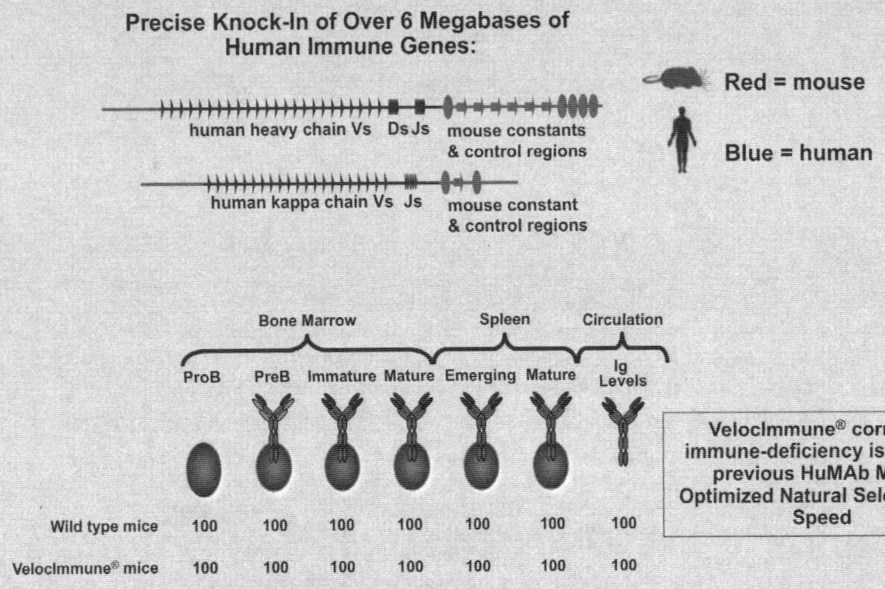

VelocImmune® corrects immune-deficiency issues of previous HuMAb Mice: Optimized Natural Selection & Speed

REGENERON
science to medicine

Humanizations in Product Development

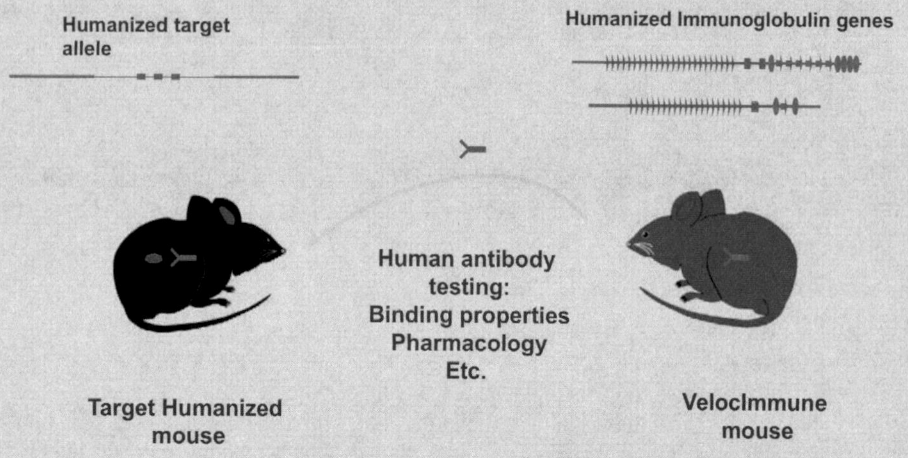

REGENERON
science to medicine™

BST: B-cell Sorting Technology

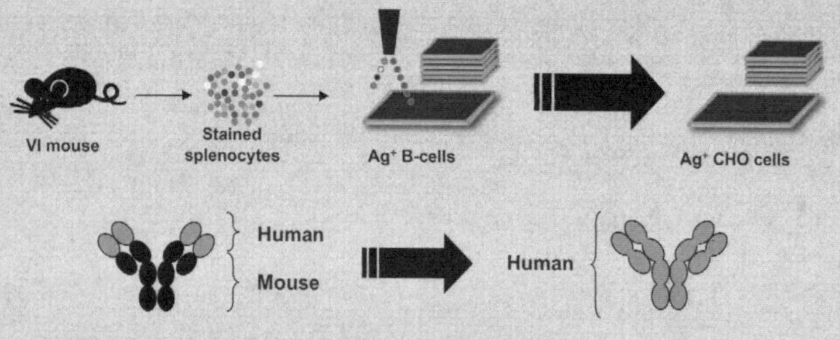

- Single B cells are isolated by flow cytometry
  - >10,000 single antigen-positive B cells
- Thousands of Ab variable domains are amplified by PCR and cloned into human IgG vectors
  - Fully automated process
- All antibody screening and characterization use fully human molecules expressed from production CHO cells
  - Production cell line available immediately upon candidate selection

REGENERON
science to medicine™

The EESYR® Locus

- **EESYR® is a locus specifically selected for desirable properties**
  - Regulated high expression of single copy recombinant genes
  - Efficient site-specific recombination of GOI
  - Insert stability
- **Locus tagged to monitor recombination**

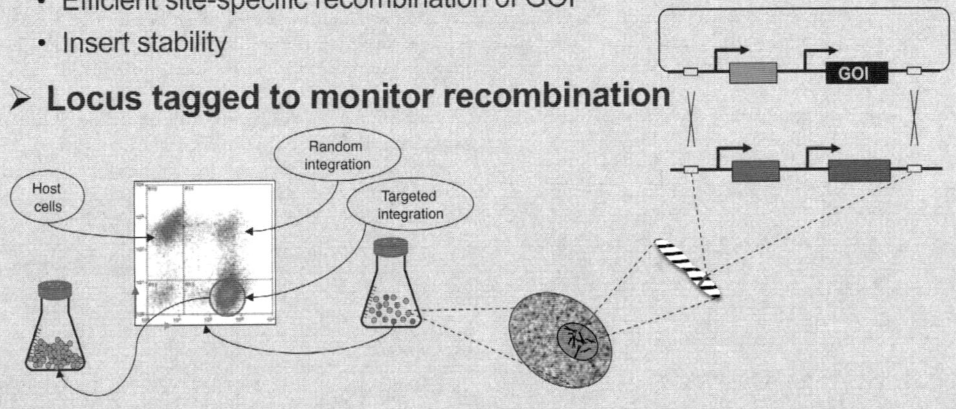

REGENERON
science to medicine®

Efficient Recombination into EESYR® Host

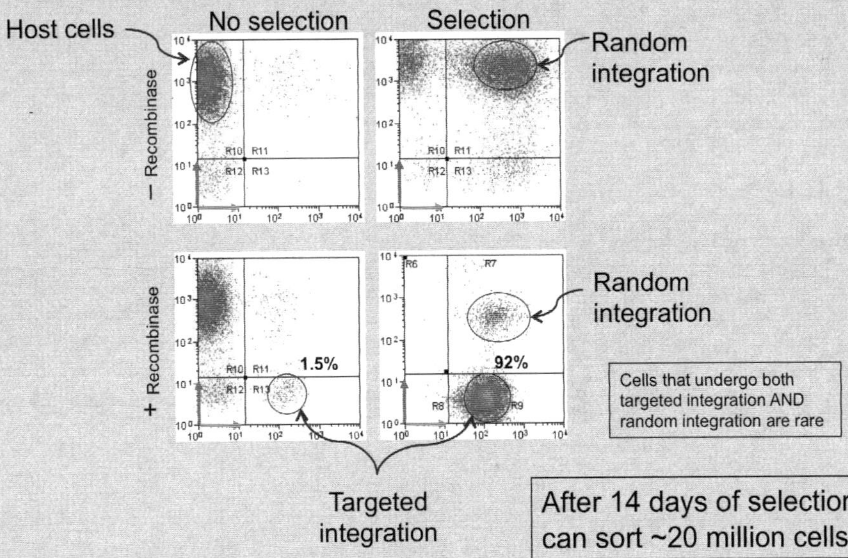

After 14 days of selection we can sort ~20 million cells

REGENERON
science to medicine®

EESYR® Cell Line System

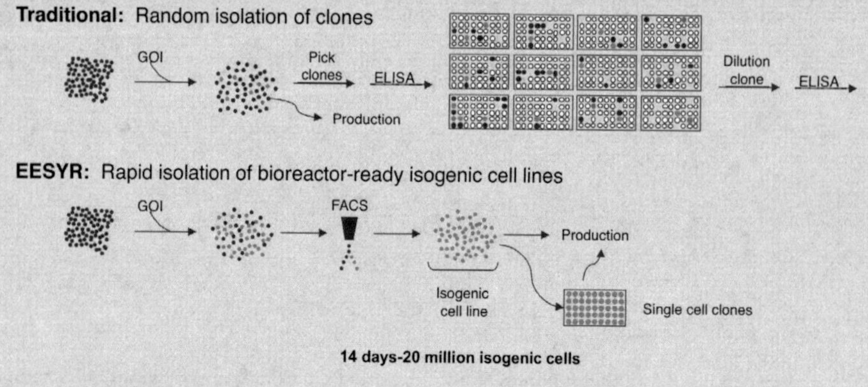

- Integrated vector, host cell, and bioprocess for the rapid construction of scalable, stable CHO cell lines
- True platform technology
  - No bioprocess optimization required
  - Predictable production levels

REGENERON
science to medicine

Integration of EESYR® Lines into VelociMab® Process

- EESYR® cells are efficiently transfected in 96-well format
- High throughput cell line construction
- Automation friendly

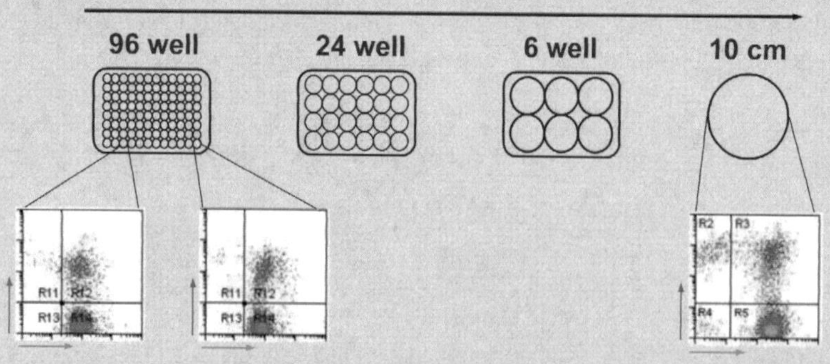

REGENERON
science to medicine

EESYR® Cell Lines Yield Consistent Clinically-Enabling Titers

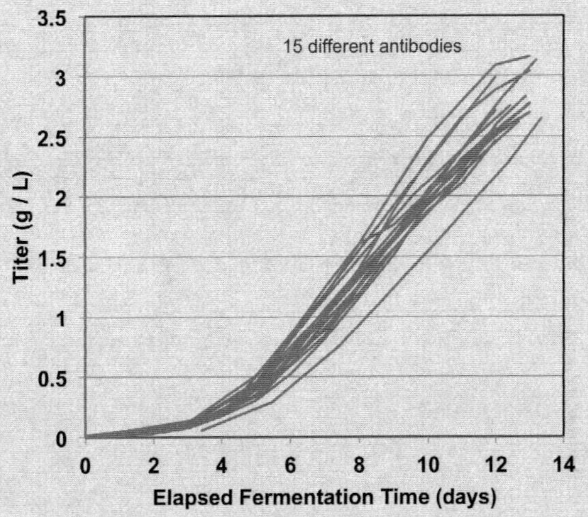

REGENERON
science to medicine®

Isogenic Cell Line Provides Early Drug Supply

- **ICLs are used for all preclinical and development activities**
  - Production of toxicology material, formulation development, purification development
- **Cloned lines used for early clinical production**
- **Minimal clone evaluation required**

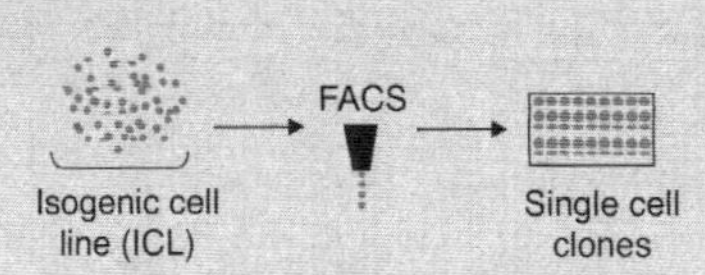

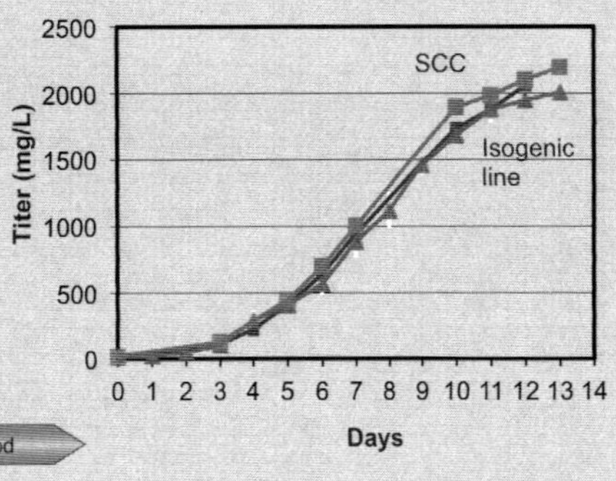

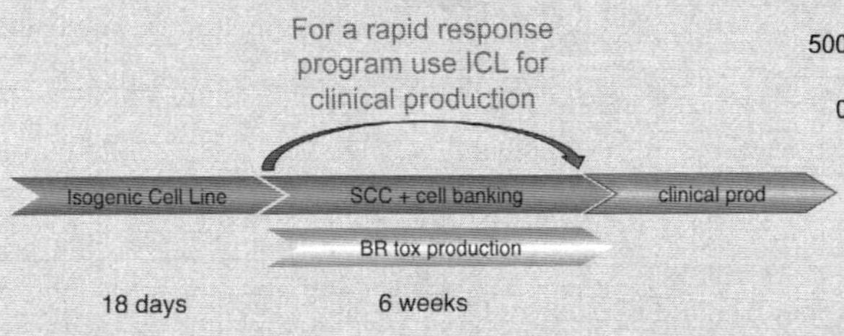

REGENERON
science to medicine®

EESYR® ICL are Suitable for Clinical Manufacturing

What is important for insuring cell substrate stability?

1. Single cell derivation
2. Maintenance of genetic homogeneity
3. Ability to evaluate homogeneity of cell population

Single cell derivation alone does not ensure a stable cell phenotype

- ICLs are genetically homogeneous, because
  - Each cell contains one copy of GOI
  - One copy of GOI is precisely integrated into a well characterized locus in the CHO genome
- Cell line drift is eliminated by repressing GOI expression throughout cell line development and banking
- The integrity of EESYR locus in individual cells can be monitored by evaluation of eGFP by flow cytometry

REGENERON
science to medicine

Clones from EESYR® Recombinant Population are Isogenic

- **Single cell clones derived from recombinant population show exceptional homogeneity**
  - One copy in a well-characterized locus
  - Consistent production levels between clones
- **Repressed GOI expression maintains genetic homogeneity**

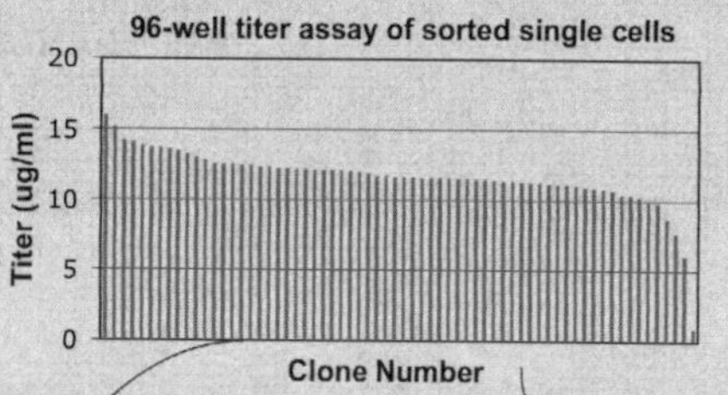

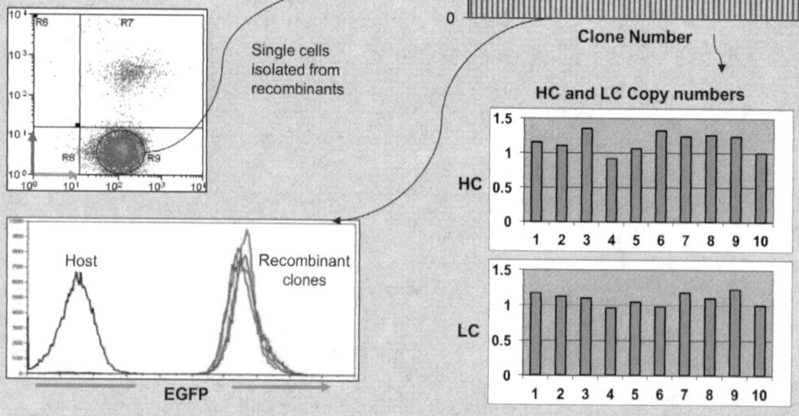

REGENERON
science to medicine

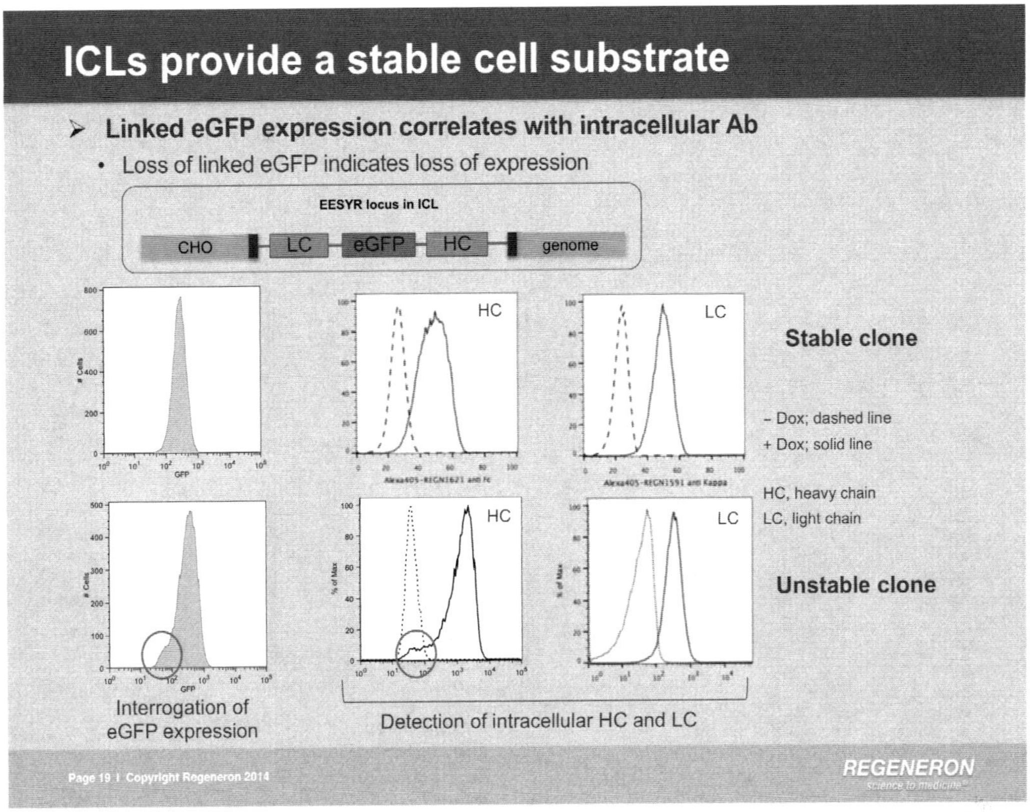
ICLs provide a stable cell substrate
Linked eGFP expression correlates with intracellular Ab
Loss of linked eGFP indicates loss of expression
EESYR locus in ICL
CHO
LC
eGFP
HC
genome
HC
LC
Stable clone
- Dox; dashed line
+ Dox; solid line
HC, heavy chain
LC, light chain
Unstable clone
Interrogation of eGFP expression
Detection of intracellular HC and LC
Page 19 I Copyright Regeneron 2014
REGENERON
science to medicine

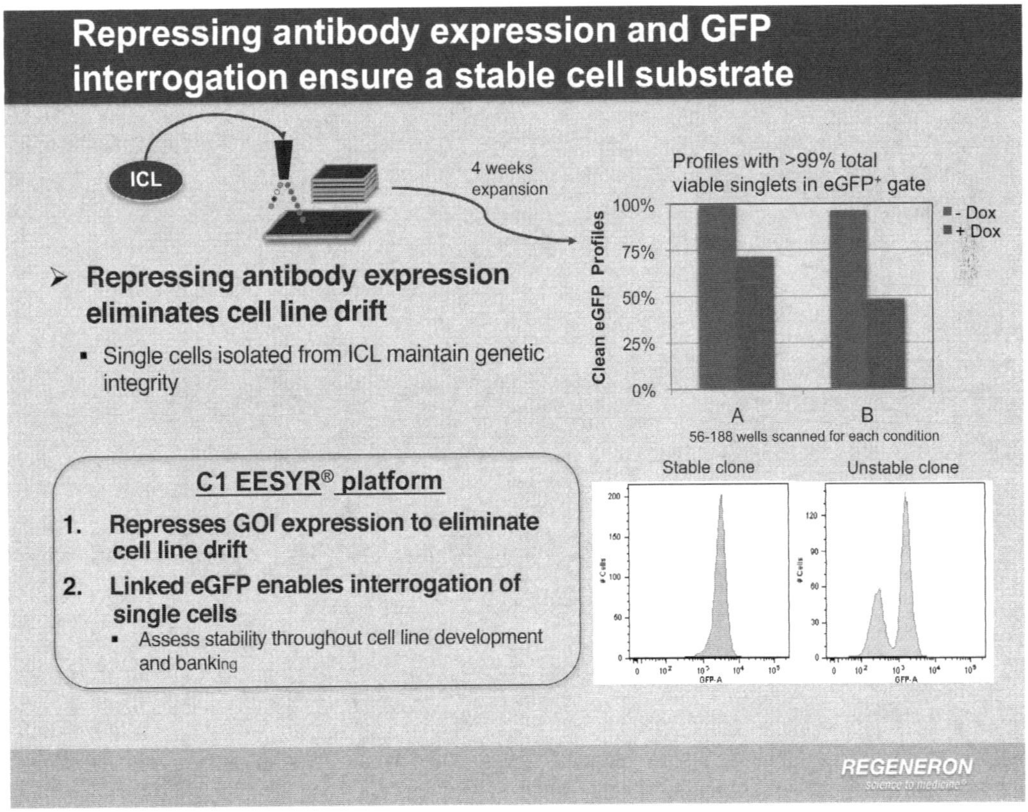
Repressing antibody expression and GFP interrogation ensure a stable cell substrate
ICL
4 weeks expansion
Repressing antibody expression eliminates cell line drift
Single cells isolated from ICL maintain genetic integrity
Profiles with >99% total viable singlets in eGFP+ gate
Clean eGFP Profiles
100%
75%
50%
25%
0%
- Dox
+ Dox
A
B
56-188 wells scanned for each condition
Stable clone
Unstable clone
C1 EESYR® platform
1. Represses GOI expression to eliminate cell line drift
2. Linked eGFP enables interrogation of single cells
Assess stability throughout cell line development and banking
REGENERON
science to medicine

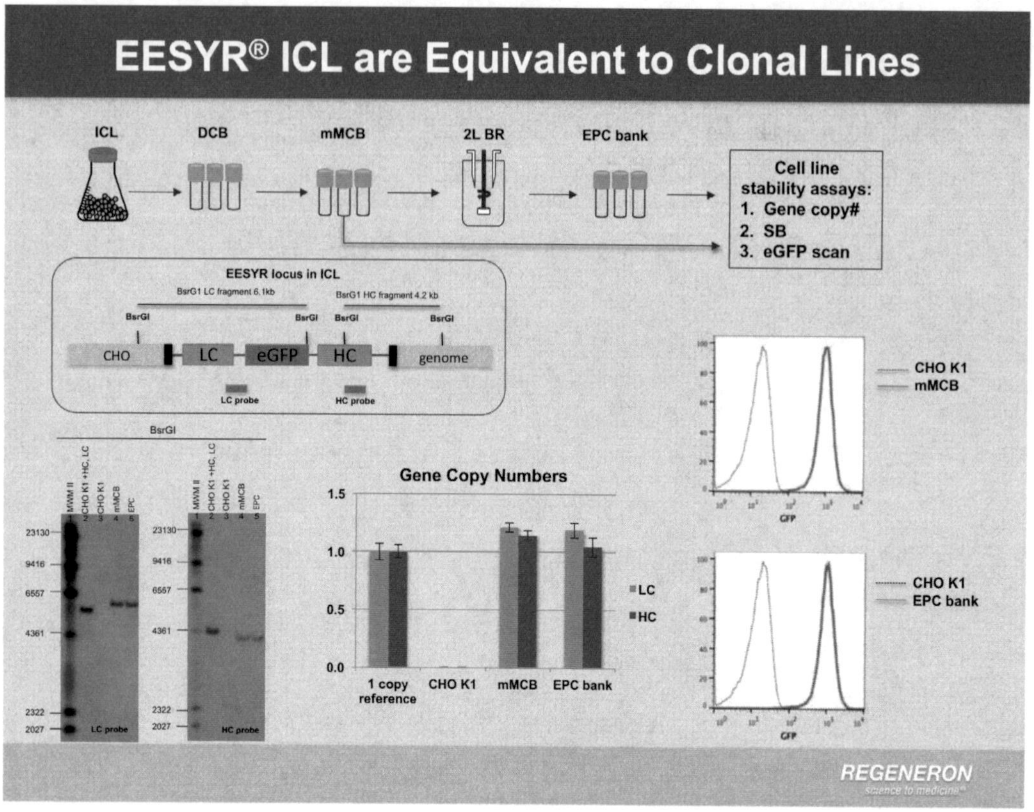
EESYR® ICL are Equivalent to Clonal Lines
ICL
DCB
mMCB
2L BR
EPC bank
Cell line stability assays:
1. Gene copy#
2. SB
3. eGFP scan
EESYR locus in ICL
BsrG1 LC fragment 6.1kb
BsrG1 HC fragment 4.2 kb
BsrGI
CHO
LC
eGFP
HC
genome
LC probe
HC probe
Gene Copy Numbers
1.5
1.0
0.5
0.0
1 copy reference
CHO K1
mMCB
EPC bank
LC
HC
CHO K1
mMCB
CHO K1
EPC bank
GFP
23130
9416
6557
4361
2322
2027
LC probe
HC probe
REGENERON
science to medicine

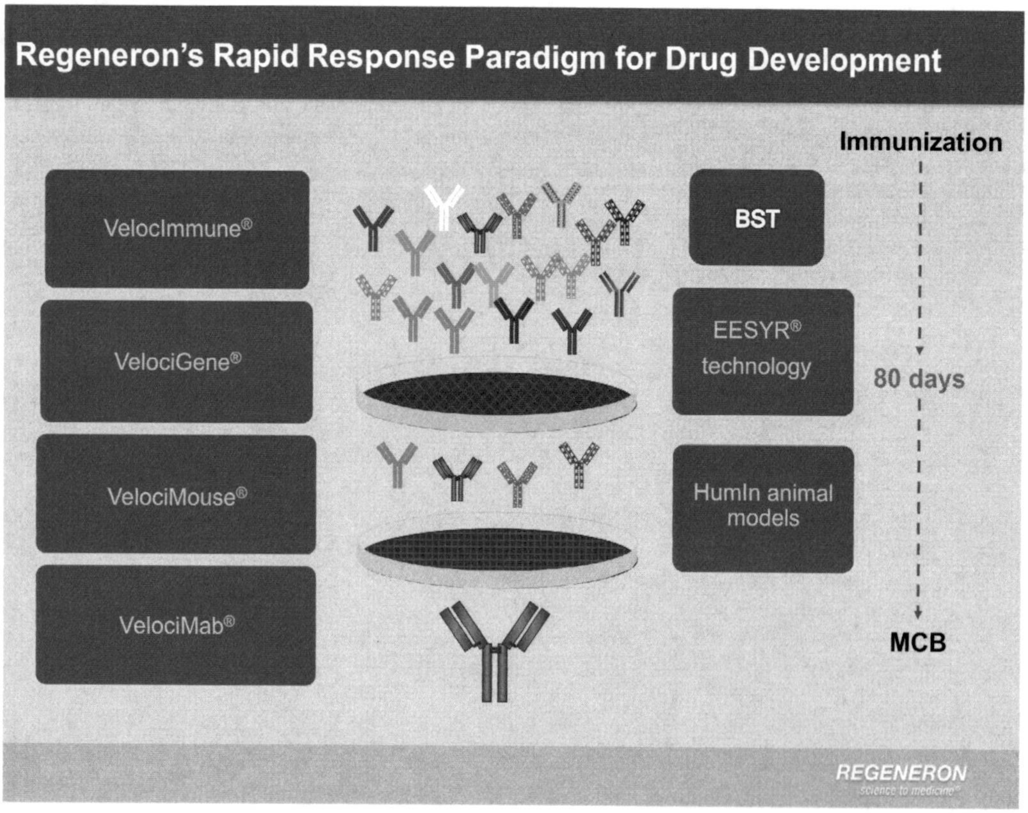
Regeneron's Rapid Response Paradigm for Drug Development
VelocImmune®
VelociGene®
VelociMouse®
VelociMab®
BST
EESYR® technology
HumIn animal models
Immunization
80 days
MCB
REGENERON
science to medicine

Regeneron's Rapid Response Paradigm: 80 Days from Immunization to MCB

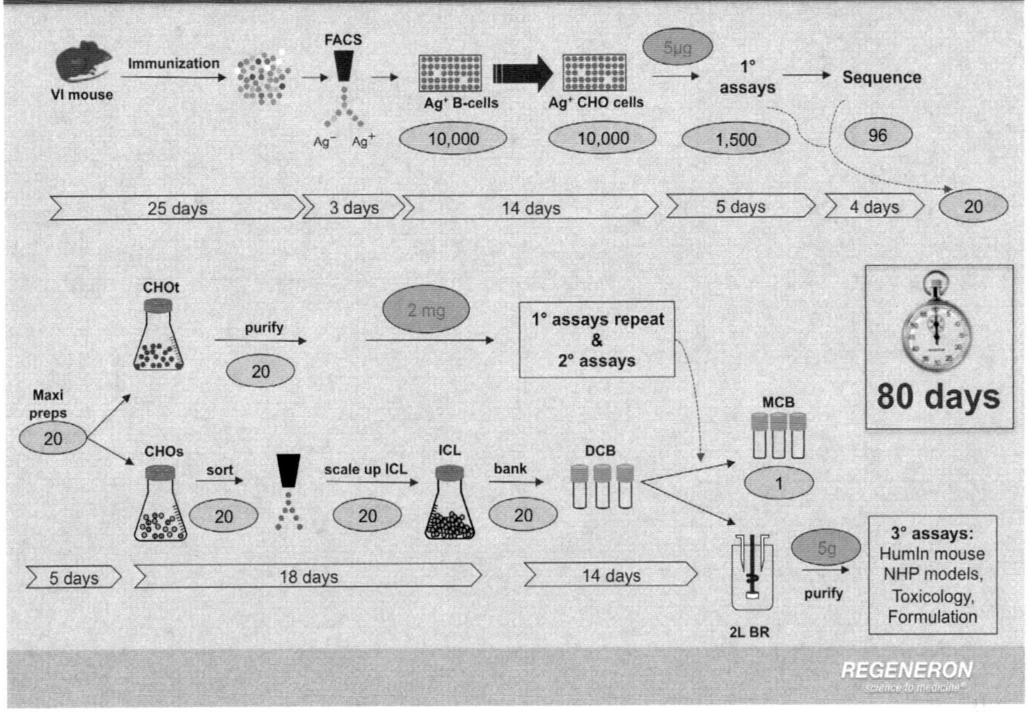

Summary

- Rapid response to infectious diseases is increasingly important
- Regeneron's technologies enable rapid isolation, development, and production of fully human mAbs to address this need
  - Cell line development can be expedited to 18 days from transfection to DCB hand off by using ICLs
  - Cell substrate stability of ICLs is equivalent to clonal cell lines

REGENERON
science to medicine

Thank you
Questions?
REGENERON
REGENERON

Towards making biologic drugs on demand

Love, J. C., Massachusetts Institute of Technology

Originally presented on: 3/2/2015 1:15:00 - 1:40:00

The current timeline for manufacturing high-quality biologic medicines and delivering them to patients typically requires many months. There are many circumstances, however, where the rapid production, release, and delivery of these drugs could address unmet medical needs, including drug shortages, emergency situations, and pandemics. This talk will present a new platform developed to produce biologic medicines on demand called Integrated and Scalable Cyto-Technology (InSCyT). This platform comprises an integrated, subliter-scale portable system for the (semi)continuous operation of fermentation, filtration of cell debris from secreted product, innovative affinity-based purification, polishing, and finishing. This program also emphasizes integrated on-line PAT and process control for QbD production and product qualification for release. Examples of the underlying technologies enabling this platform will be presented, along with a demonstration of the integrated operation to produce high-quality protein therapeutics in less than 48 hours.

Beyond the applications for rapid manufacturing of biologics at point-of-care, the speed of production enabled by automation and integration highlights other opportunities for drug development and conventional manufacturing. These include modeling complete manufacturing processes during process development, rapidly creating materials for preclinical and early-stage clinical testing, and manufacturing therapies for orphan diseases, among others.

Development of a Device Platform to Speed Products to Clinic and Market
Peeples, R. E., Pfizer
Originally presented on: 3/2/2015 1:40:00 - 2:05:00

An increasing number of therapies using biologics are requiring home-administration by the patient or a caregiver. To assist in the proper administration of these medications and ensure proper compliance, administration devices such as autoinjectors are being more commonly used. In the past, due to the development time needed for a device and early financial commitment, many products were first marketed in a vial or syringe and a device was introduced as a next generation dosage form.

In order to speed the process to provide a device for the clinic and market, a platform of devices was developed that can be applied quickly and cost-effectively. This presentation will describe the reasons for using a platform approach and lessons learned.

Continuous vs. Batch Bioprocessing

Navigating in Uncertain Waters to Transform Pharmaceutical Manufacturing

Charles L. Cooney
Robert T. Haslam Professor of Chemical Engineering, Emeritus
MIT Cambridge MA

Accelerating Biopharmaceutical Development
Scottsdale, AZ
March 3-5, 2015

Historical Perspective to Meet Product Demand – Industrial Evolution

- **Industrial Intensification**
  - Steel
  - Chemicals
  - Food
  - Pharmaceuticals
- **Driven by**
  - Need to reduce cost
  - Desire to decentralize manufacturing
  - Robust quality
  - Flexibility to meet product demand

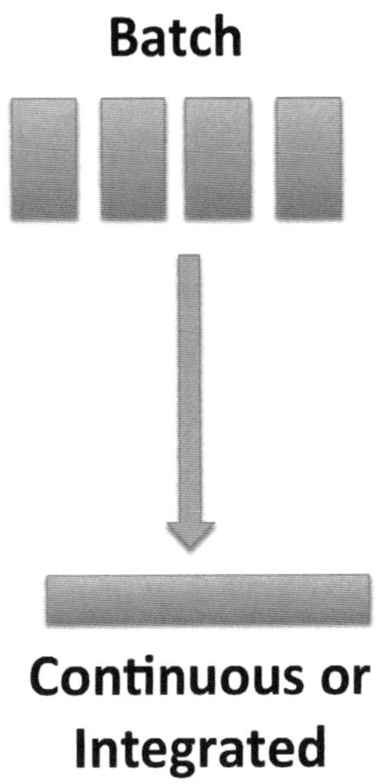

Defining Continuous

- A ***unit operation is continuous*** if it is capable of processing a continuous flow input for prolonged periods of time
  - minimal internal hold volume
  - output can be continuous or discretized in small packets produced in a cyclic manner.
- A ***process is continuous*** if it is composed of integrated continuous unit operations with zero or minimal hold volume in between.

In continuous operation, relevant properties (CQAs) are under control and the product meets its TPP criteria

Novartis-MIT Center for Continuous Manufacturing
Department of Chemical Engineering, MIT

MIT- *Professors*

- Paul Barton: Chem E
- Richard Braatz: Chem E
- Steve Buchwald: Chemistry
- Jung-Hoon Chun: Mech E
- Charles Cooney: Chem E
- Timothy Jamison: Chemistry
- Klavs Jensen: Chem E
- Allan Myerson: Chem E
- Greg Rutledge: Chem E
- Alex Slocum: Mech E
- Bernhardt Trout: Chem E

Novartis

- Juan Andreas
  Head Global Techn. Operations
- Markus Krumme
  Head Continuous Unit
- Steffen Lang
  Head Global Techn. R&D
- Berthold Schenkel
  Chemical Development
- Norbert Rasenack/Stefan Hirsch
  Pharmaceutical Development

+ students and Post-docs

LEAN Manufacturing: improving asset utilization

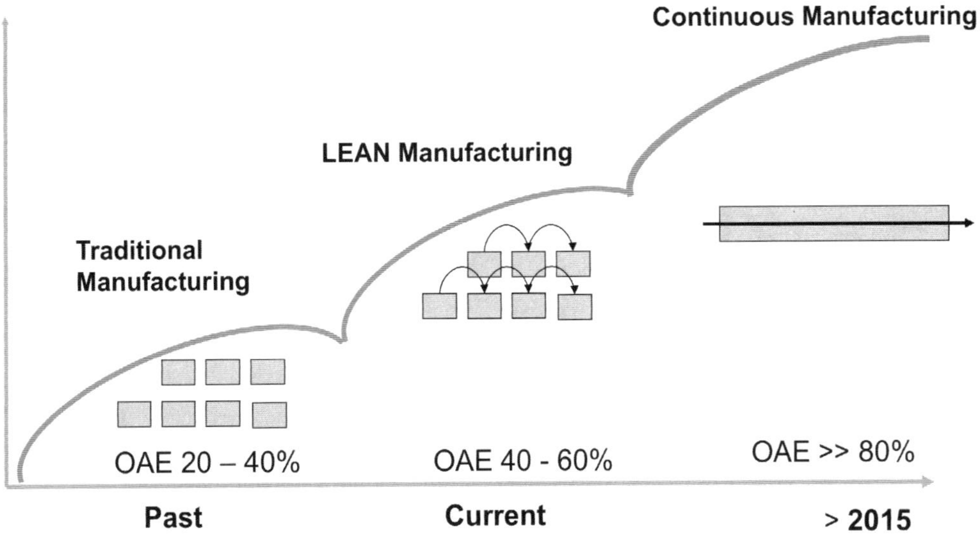

OAE: Operational Asset Effectiveness

MIT NOVARTIS

Some of the Innovations: "Converging Technology"

- Solvent free synthesis
- New, fast chemistry (cannot do in batch)
- Eliminate non-value add unit operations (blending, granulation, milling) to reduce processing time and increase yield
- Integrate new unit processes:
  - filtration and drying
  - Mold extrusion
  - Continuous crystallization
  - Solvent free coating
- Route for QbD for CM process

MIT NOVARTIS

Lessons Learned

- Rethink the process and design for continuous operation
- Implement only value added steps avoiding "corrective" actions
- Recognize convergence of technology
- Utilize QbD to drive mechanistic understanding of process science
- Minimize variability of raw materials
- Scale out, not up, with time and parallel operations

7 |

MIT NOVARTIS

Where are we Today?

- Forty MAB products in the market (Mostly US & Europe) and sales at ~15%/yr
- Capacity in place to make >50 ton MAB/year
- Clinical trials dominated by biologics
- Large unmet medical needs by both disease <u>and</u> geography
- Patent expiration and market exclusivity expiring
- Biosimiliars beginning to appear

Where will we be in 2025?

- A clinical pipeline still dominated by biological molecules?
- Biosimilars readily available and interchangeable?
- A future where a blockbuster drug is selling to a billion people instead of a billion dollars?
- What will it take to get us there?
  - One plant supplying the global market (stainless steel "six packs")?
  - Individual plants for each patient (cell therapy)?
  - Personalized and Precision medicine (gene therapy)?
  - Multiple & distributed facilities dedicated to a individual markets?

Continuous Biomanufacturing is Not New

- Continuous culture – 1950's
- Continuous extraction of penicillin 1950's
- Continuous chromatography 1950-60's
- Extended fed-batch (aka Perfusion) for penicillin production 1960's
- Tangential flow filtration 1960's
- Continuous crystallization for AA's – 1960's
- ...

Today's Platform for MAB's

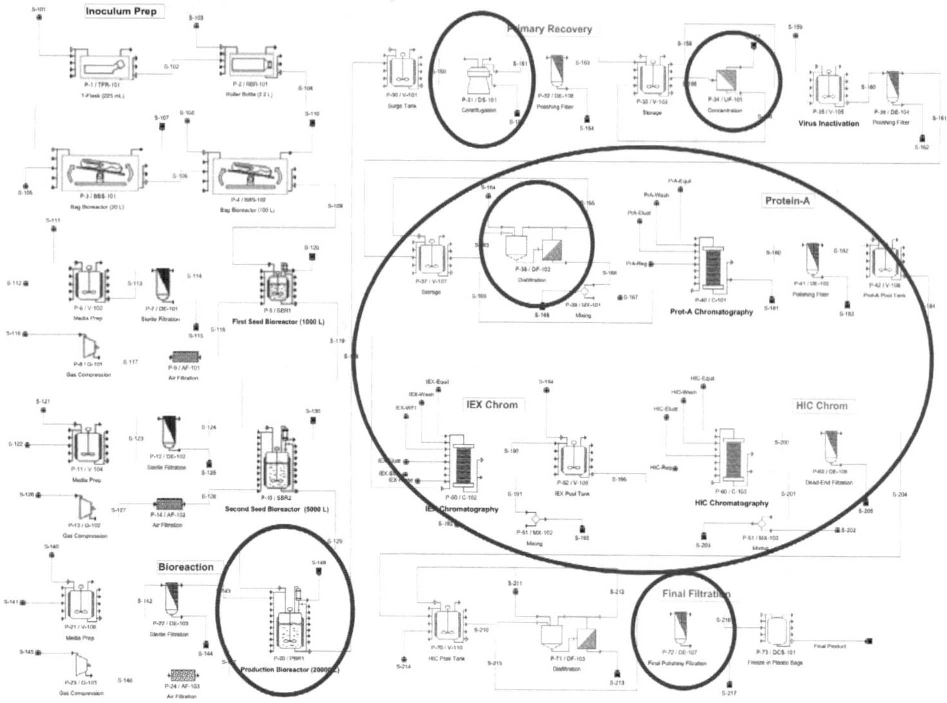

International Symposium on Continuous Manufacturing of Pharmaceuticals MIT, May 20-21, 2014

White Paper on Continuous Bioprocessing

Authors and first reviewers:

Charles Cooney & Konstantin Konstantinov

Tim Charlebois, Thomas Daszkowski, Uwe Gottschalk, Chetan Goudar, Sa Ho, Wei-Shou Hu, Chris Hwang, Tim Johnson, Karol Lacki, Bob Mattaliano, Massimo Morbidelli, Thomas Mueller-Spaeth, Jens Vogel, Veena Warikoo, Bill Whitford, Andrew Zydney

With Feedback from many colleagues

White Paper on Continuous Bioprocessing JPS 2014

International Symposium on Continuous Manufacturing of Pharmaceuticals MIT, May 20-21, 2014

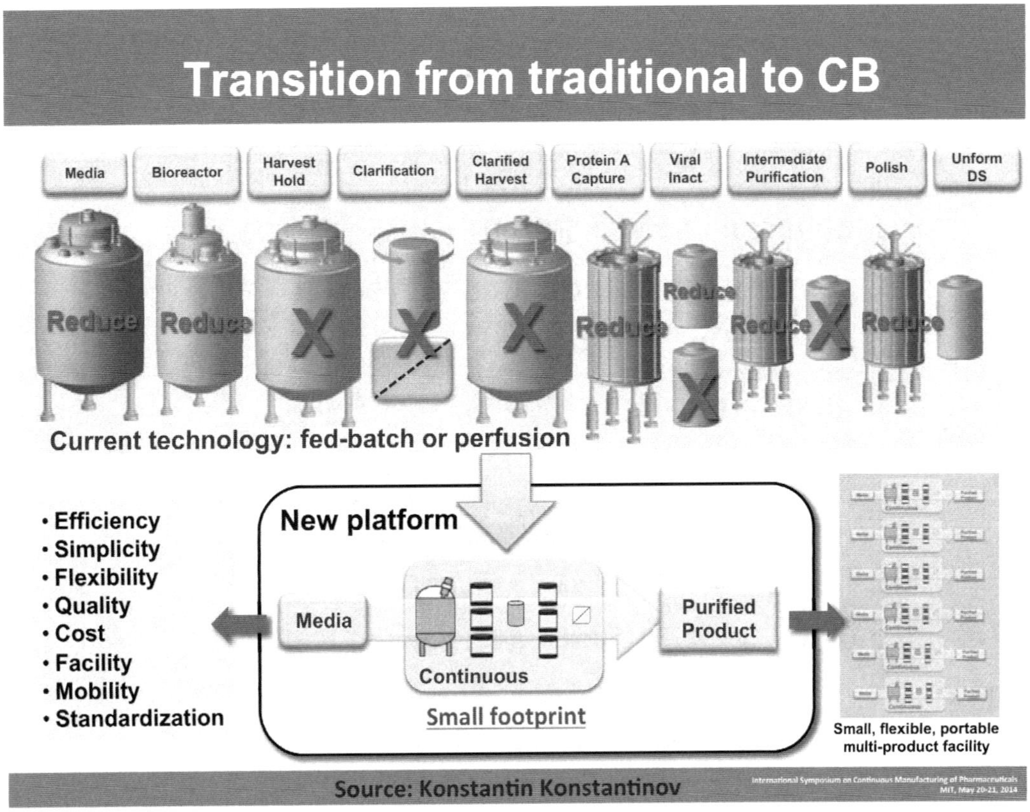

Benefits from Continuous Biologicals Manufacturing

Feature	Benefit
Flexibility in scaling by time and parallelization	Fast process development and flexible production
Smaller equipment and process footprint	Reduction in capital expense and construction/validation time
Short process residence time	Reduction in degradation; improved yield and quality
Flexibility in scheduling	Rapid response to demand
Continuous Improvement	Driving cost from >$100 → <$10/gm

Robust Quality

Flexibility **Cost/Profit**

Technology Needs - Upstream

- Stable cell lines (2-3 months)
- Operating on defined animal-free media
- Cell retention (perfusion)
- ✓ High cell density > 100 x 10^6 cell/ml
- ✓ High productivity (1-2 reactor vol/day)
- Control (O_2, CO_2, feeding) at high density and productivity
- Small bioreactors – single- *vs*. Multi-use

International Symposium on Continuous Manufacturing
of Pharmaceuticals MIT, May 20-21, 2014

Technology Needs Downstream

- ✓ Interface with perfusion culture and DSP
- ✓ Product capture
- ✓ Product purification
- Viral inactivation & removal
- UF/DF and in-line mixing
- PAT to assure CQA of product

International Symposium on Continuous Manufacturing
of Pharmaceuticals MIT, May 20-21, 2014

Process Issues

- Platform technology beyond MAB's – unstable proteins
- ✓ Standardization across technologies
- Global process control to support CQAs
- ✓ Defining *steady variation* within acceptable bounds
- Cleaning standards
- ✓ Incorporating & integrating single-use technologies

International Symposium on Continuous Manufacturing
of Pharmaceuticals MIT, May 20-21, 2014

Regulatory Issues

- Lot size may be defined by mass <u>or</u> time
- Continuous manufacturing is within guidelines
- Performance defined as outcome of control
- ✓ Manage a product in multiple markets
- ✓ Align vision, review and inspection

International Symposium on Continuous Manufacturing
of Pharmaceuticals MIT, May 20-21, 2014

Business Issues

- Compelling business model
- New organizational structure
- Management comfort in context of *perceived* regulatory constraints
- Large installed batch capacity
- Holistic view of supply chain (raw materials, technology, customers)
- Managing demand variability with flexibility

International Symposium on Continuous Manufacturing
of Pharmaceuticals MIT, May 20-21, 2014

A Roadmap with Multiple Process Configurations

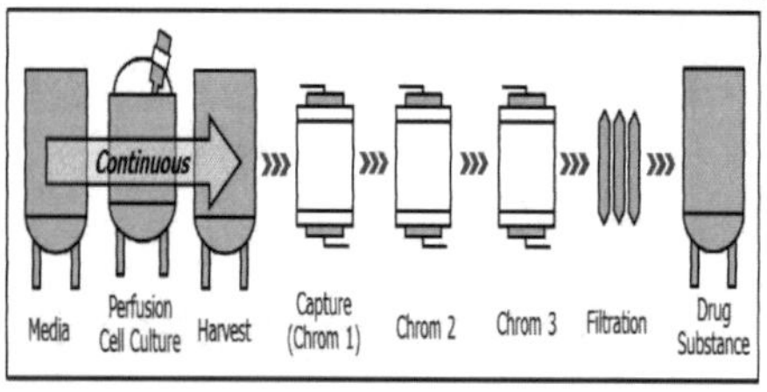

Example 1: Continuous upstream (perfusion), batch downstream

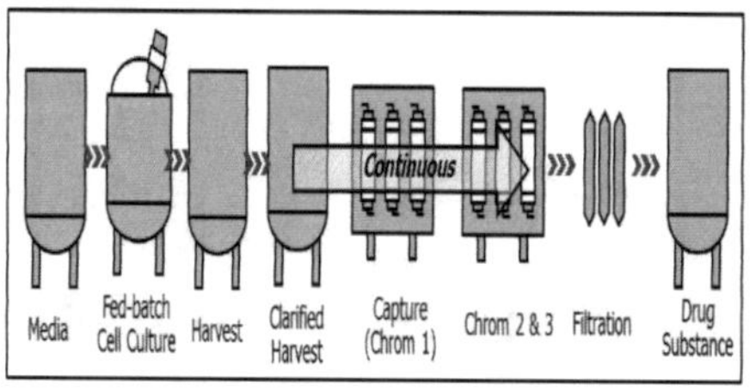

Example 2: Batch upstream, continuous downstream

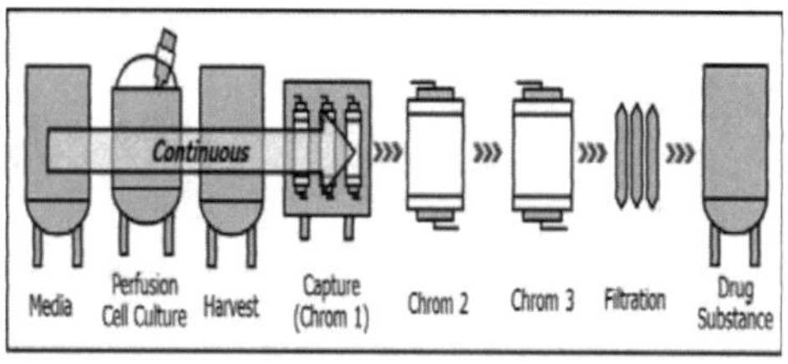

Example 3: Continuous upstream + capture, batch downstream

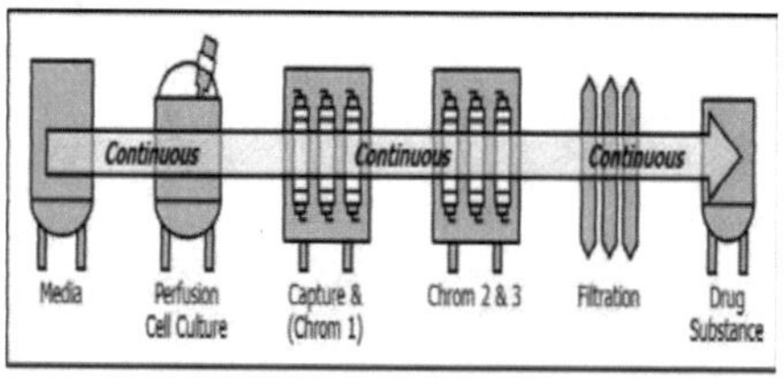

Example 4: Continuous upstream and downstream

Source: Konstantinov and Cooney, JPS 2014

Continuous Bioprocessing – The Opportunity

- Manufacturing for the clinic and launch
- Scaling to meet market demand in time and place
- Reducing time & cost of development
- Reducing the cost of production over life cycle
- Driving cost from >$100/gm → <$10/gm
- Assuring quality in real time
- Managing the integrity and security of complex supply chains

Does Pfizer Need a Low Cost Low Volume Manufacturing Option?

Salm, J., Pfizer, Inc
Crowley, T., Pfizer, Inc
Shang, T., Pfizer, Inc
Barry, A., Pfizer, Inc
Murray, K., Pfizer, Inc
Originally presented on: 3/3/2015 0:25:00 - 0:50:00

A Low Cost, Low Volume (LCLV) approach to clinical manufacturing has previously been proposed[1,2]. The goal of this approach was to provide Pfizer research organizations with an option for initiating first-in-human studies in the clinic for a reduced initial investment. The offer provided 150 g of drug substance for regulatory toxicology studies and approximately 1000 drug product vials for approximately half the cost compared with Pfizer's traditional approach for monoclonal antibody (mAb) programs.

This LCLV approach was recently implemented for a Phase 1 mAb program. The program used a precision medicine approach targeting a new pathway in a specific population for a disease with insufficient treatment options. The goal of the initial clinical study was to demonstrate proof of mechanism. Since the material demands were thought to be low and the cost of entering the clinic was limiting, the program was ideally suited for LCLV.

During LCLV implementation, certain aspects of LCLV were found to be very enabling. Process development was successfully eliminated which significantly reduced the cost to the program. Several regulatory elements were tested including a modular viral clearance approach and reduced DS testing and stability. However, other aspects of LCLV were found to be too restrictive for the program including the amount of regulatory toxicology and clinical trial material that was initially generated.

This talk will discuss the first implementation of the LCLV process. The assumptions of the original approach will be reassessed in light of this first clinical program. Specific aspects of the process that have been successful as well as future areas of opportunity will be discussed. An improved version of LCLV paradigm will be proposed along with the potential implications and value LCLV might have on Pfizer's overall development portfolio.

1. Salm, J., Shang, T., Crowley, T., Coffman, J., Moxham, J., (2013, February). Increasing Access to the Clinic with a Low Cost, Low Volume Manufacturing Process, presented at *SBE's 3rd International Conference on Accelerating Biopharmaceutical Development*, Coronado Island, CA.

2. Shang, T., Salm, J., Porter, T., Charlebois, T. (2014, May). Accelerating clinical evaluation of biotherapeutics: can cost, speed and throughput be improved without compromising quality? Presented at *CMC Strategy Forum Europe*, Sorrento, Italy.

UCL

Ultra scale-down (USD) technologies
for a scaleable and informed biomanufacturing

AccBio2015
Scottsdale, AZ
1- 4th Mar 2015

Andrea Rayat, PhD
Alex Chatel, Prof Gary Lye, Prof Mike Hoare

Department of Biochemical Engineering
University College London

mbi modular training for the bioprocess industries

UCL ENGINEERING
Change the world

The Queen's Anniversary Prizes for Higher and Further Education 2013

UCL

Typical development timeline

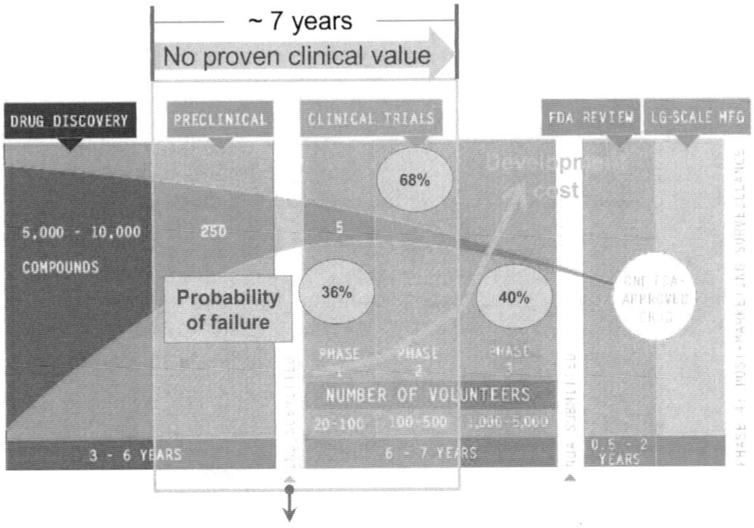

Assess multiple candidates earlier for manufacturability

Some barriers for effective and *efficient* biopharmaceutical development

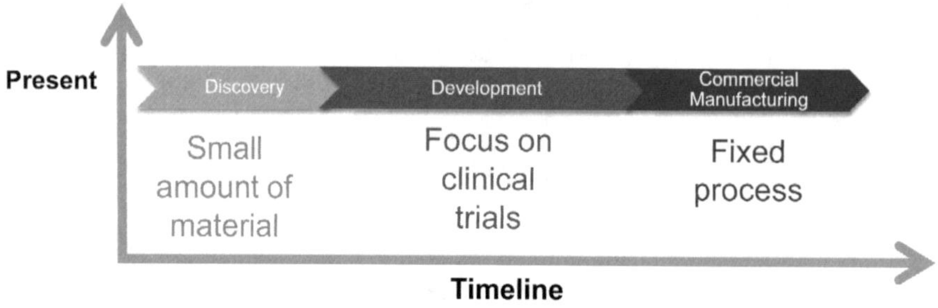

- Drug discovery is key activity; efficacy is main concern
- Manufacturability assessment happens towards the end of the process
- Limited scope for designing manufacturability into product → impact on process design
- Risk in scale up → cost of scale up

Towards an effective and *efficient* biopharmaceutical development

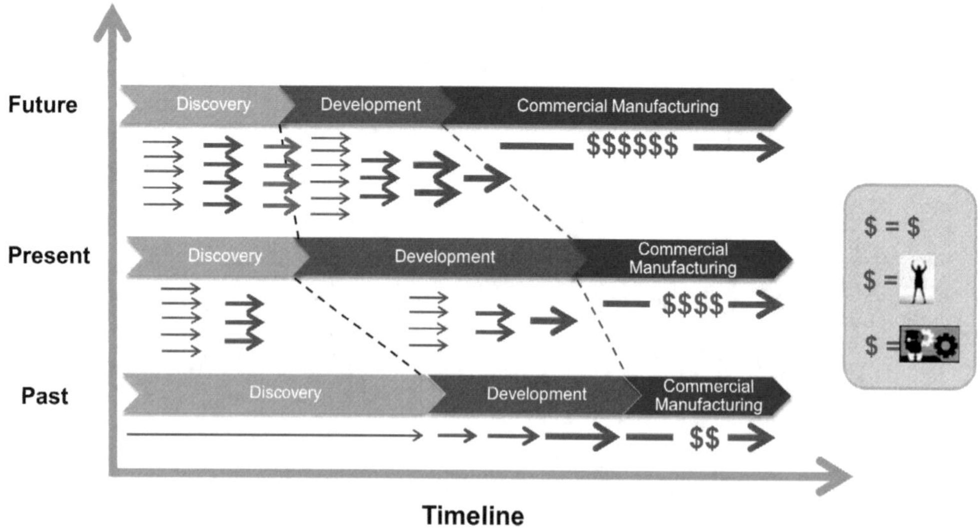

UCL

What costs?

- Cost of R&D: ~~$1B~~ $4B - $11B/approved drug
  - Cost of clinical trials at the high end is ~$100M
- Cost of manufacturing is1-5% of sales price
- Cost of capital
- Profits
- Opportunity cost

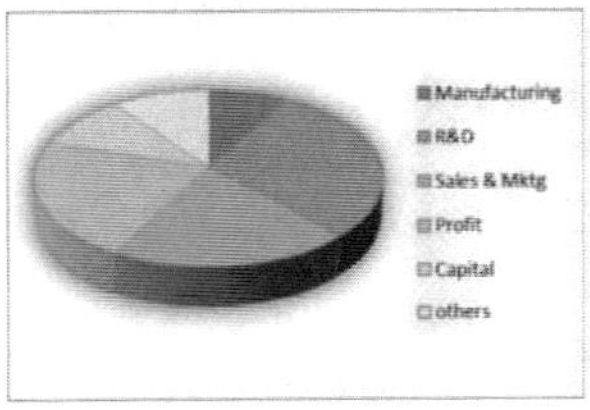

Opportunity for cost reduction greatest in R&D!
How to turn opportunity cost to gains?

Sources: *Forbes 10/2012; Kelley 2009*

UCL

Thoughts For Informing Biomanufacturing

- Assess manufacturability early
- Consider whole process sequence
- Identify interactions early and exploit them
- Quickly discard unfeasible products/processes

✓ Understand and mimic large scale process
✓ De-risk scale-up
✓ Reduce cost of manufacturing-related investigations

Platforms for informing biomanufacturing

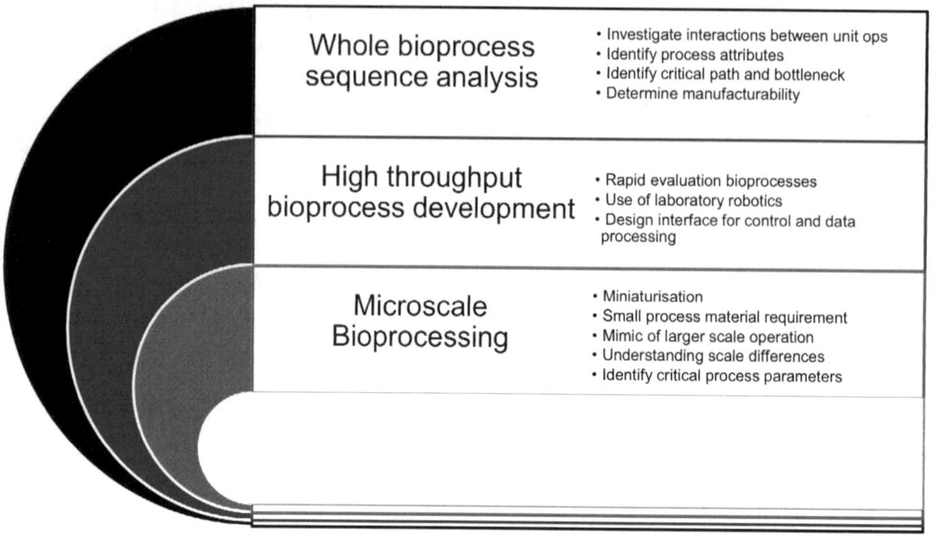

[Rayat (2011) *UCL*]

Understand and Mimic the Final Production Scale as Closely as Possible

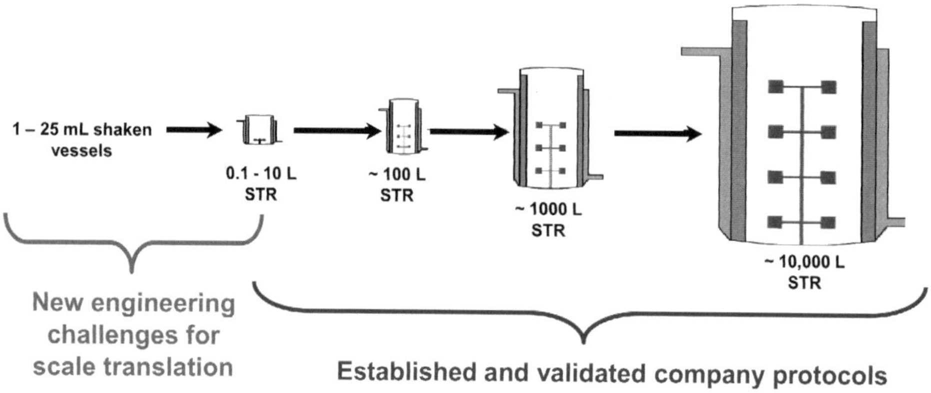

[Nienow (2006) *Cytotechnology*, **50:** 9-33]

Evolution of the use of microwell systems for High Throughput Bioprocess Development

STRs to Non-STRs | New Expression Systems | Gas supply and controls

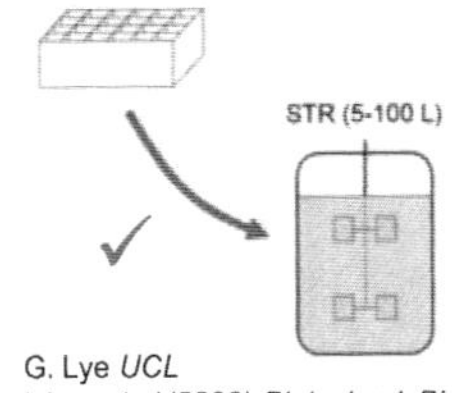

G. Lye *UCL*
Islam *et al* (2008) *Biotechnol. Bioeng.*

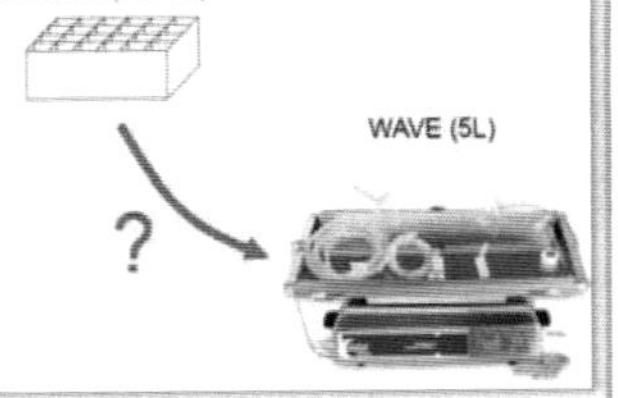

T. Mukhopadhyay UCL
Guy *et. al* 2013 *Human Gene Therapy Methods*

Fed-Batch GS-CHO* Cultures in Shaken and Stirred Bioreactors at Matched t_m Values

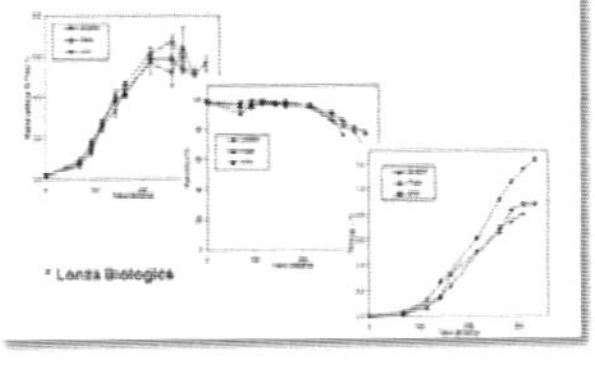

Mammalian systems more sensitive to interacting parameters: engineering, control, aeration etc

UCL EngD Project

Shaken Micro24 Bioreactor: Improved Control for Fed-Batch Mammalian Cell Culture

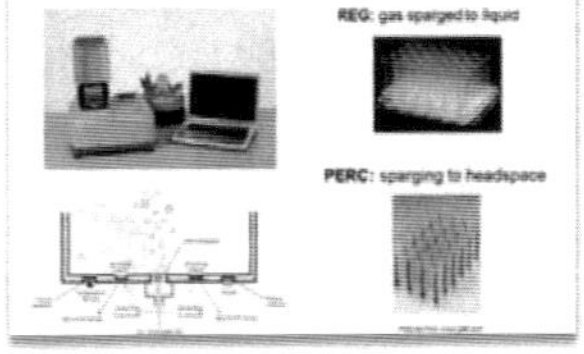

REG Plates Provide Best Indication of Stirred and Aerated Bioreactor Performance

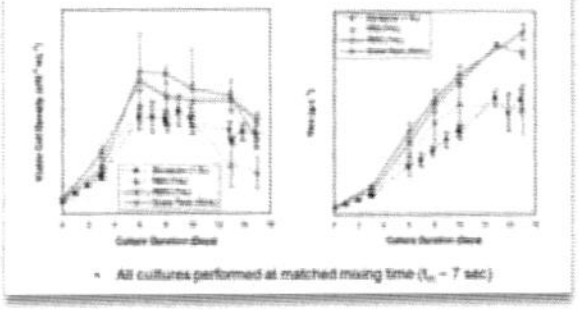

G. Lye *UCL*
Betts *et. al* 2015 *Biotech Bioeng (In prep.)*

Dimensions of process development

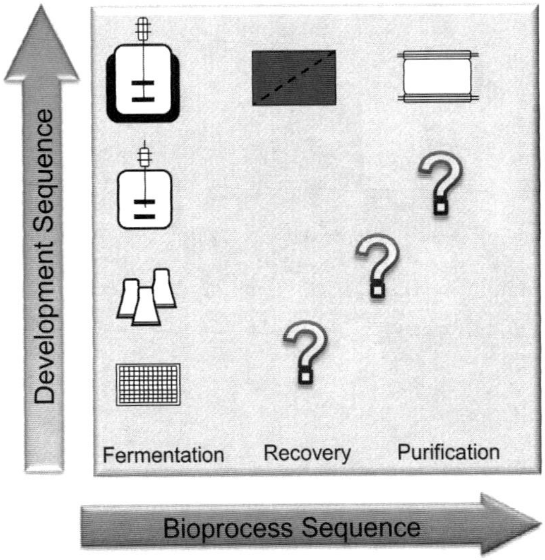

Obtain process design information early

✓ low cost initial studies

✓ acceptable investment

- small amount of material
- material quality representative of process stream
- enough material to be tested in later stages

The same considerations apply to DSP operations

What is Ultra Scale-Down (USD)?

- USD is a platform technology to help speed up the process development time
- Devices specifically designed to mimic engineering environment found in bioprocessing unit operations
- Based on thorough understanding of engineering principles governing industrial-scale bioprocessing

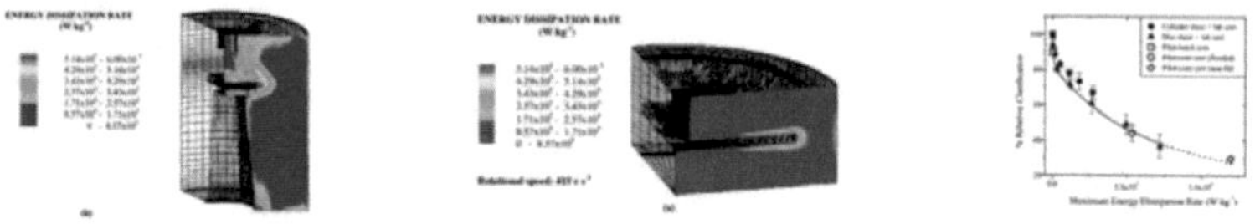

Boychyn M, et al. 2001. Characterisation of flow intensity in continuous centrifuges for the development of laboratory mimics. Chem Eng Sci 56:4759–4770.

Case study 1: USD filtration

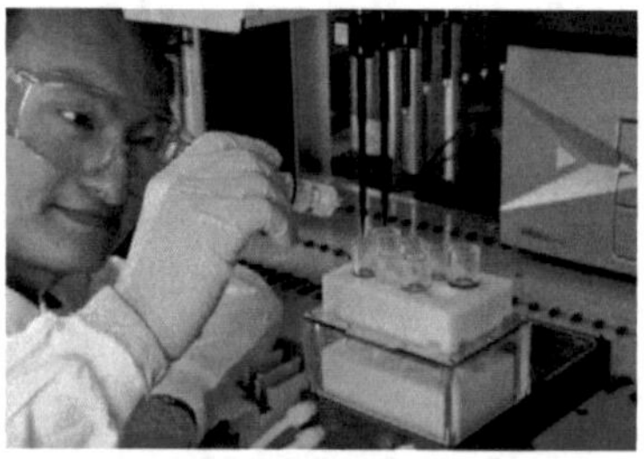

How will my process material interact with filters?

How will varying broth conditions affect filtration performance?

Initial work: Microfiltration

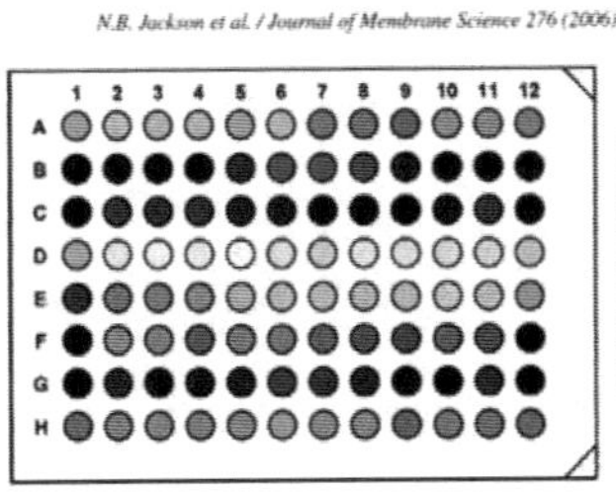

Multi-well screen filter plate

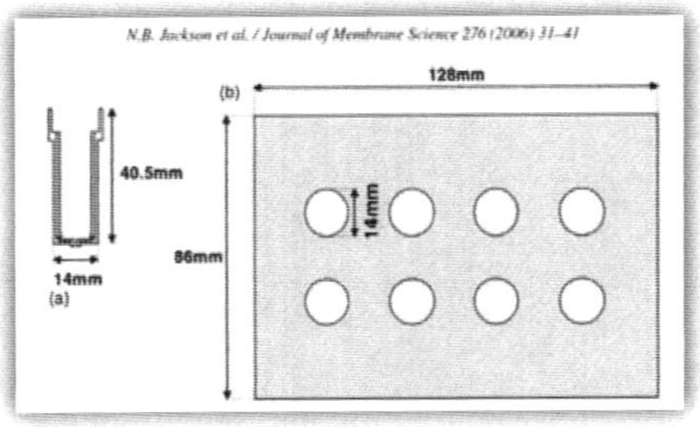

Microscale filtration plate + technique

- ➢ Up to 1.5mL material per filter insert
- ➢ Better reproducibility
- ➢ Enable study of multiple filter on the same plate
- ➢ Enable study of impact of previous step using small amount of starting material

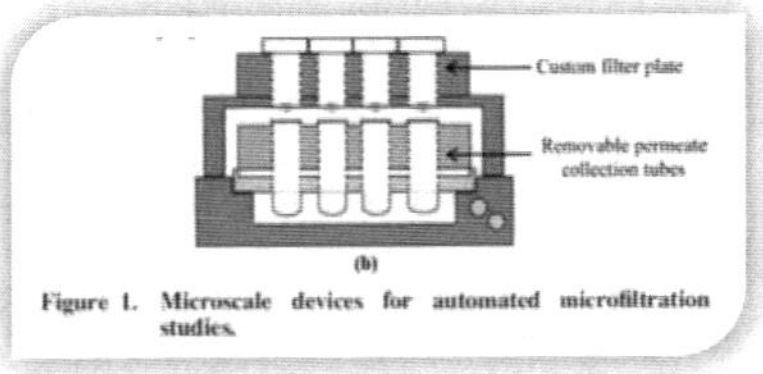

Figure 1. Microscale devices for automated microfiltration studies.

Jackson et al. Journal of Membrane Science 276 (2006) 31–41
Rayat et al. Biotechnol. Prog., 2010, Vol. 26, No. 5

Demonstrated linked automated microscale process (extraction + filtration)

Cells from fermentor →

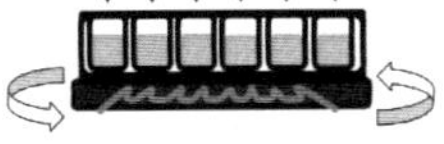

Extraction

Various extraction conditions

→

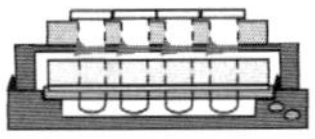

Dead-end filtration

Impact of extraction conditions on filterability ?

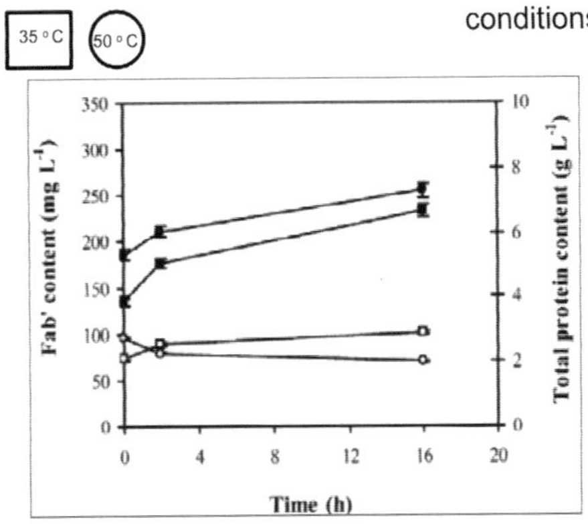

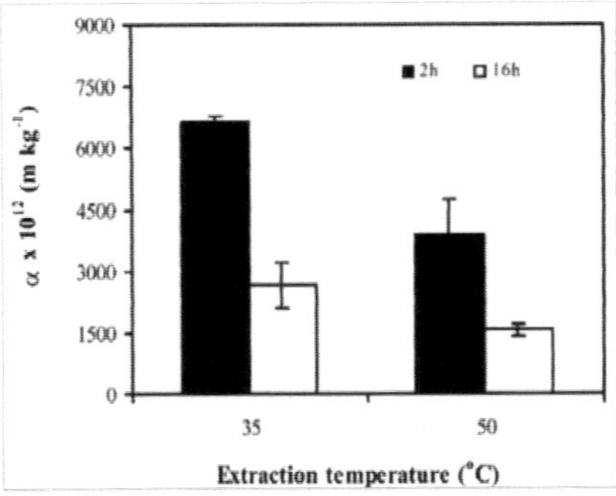

Rayat et al. Biotechnol. Prog., 2010, Vol. 26, No. 5

Current system: USD Filtration

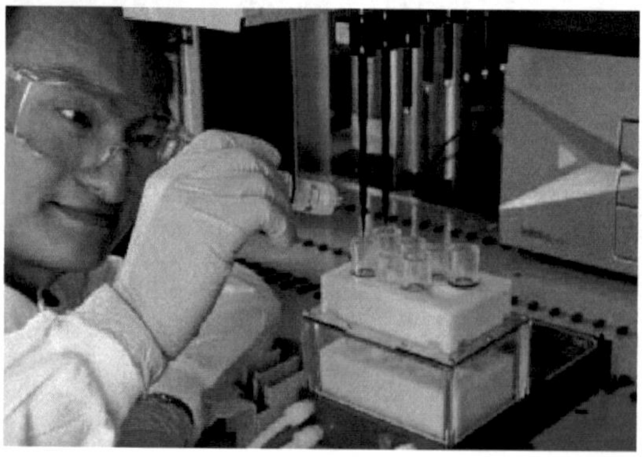

- 4mL material per filter insert
- Added automation: Liquid detection for volume measurement vs time
- Variables: pressure, filter media,
- Up to 16 simultaneous experiments
- Prepare representative material using USD shear to mimic shear in previous step (eg bioreactors or homogeniser)

USD filtration – how will my process material interact with filters?

Material:

- (clarified) cell broth expressing therapeutic protein
- USD Depth Filtration platform (on TECAN liquid-handling robot)
- Filtration media of choice

USD: **0.28 cm²**

Result: Successful prediction of process-scale filter capacity

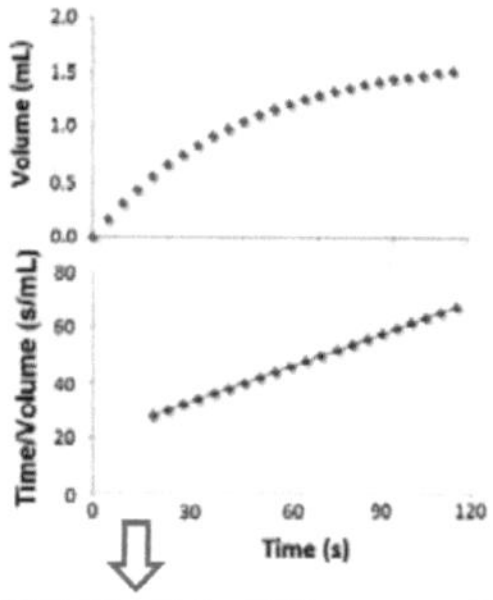

USD predicted filter capacity
V_{max} = **95 L/m²**

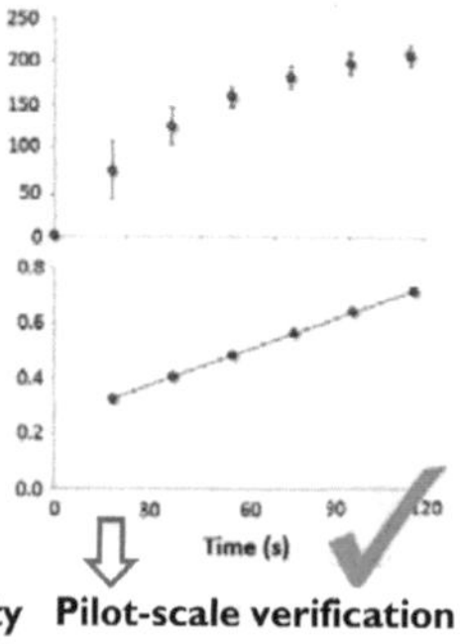

Pilot-scale verification
V_{max} = **103 L/m²**

Pilot-scale: **25 cm²**

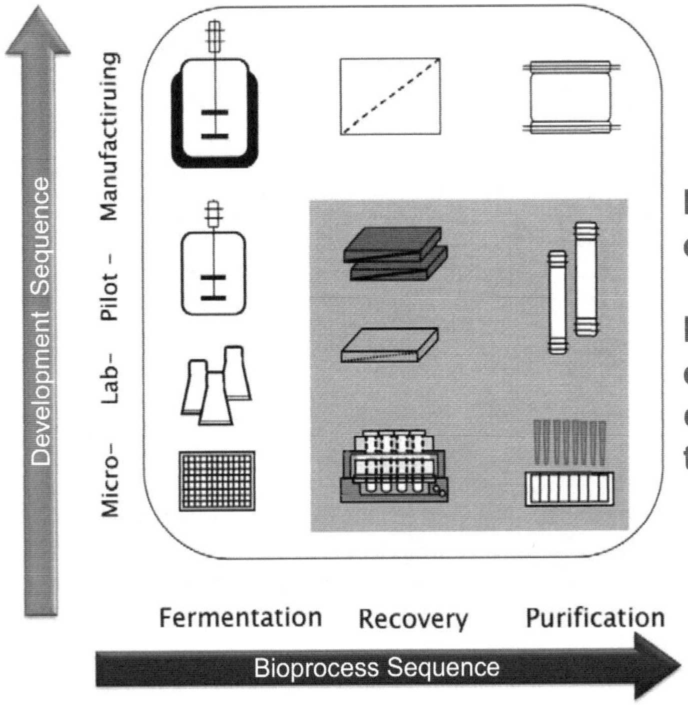

How about centrifugation?

Industrial equipment design difficult to scale-down!

Case study 2: USD Centrifugation

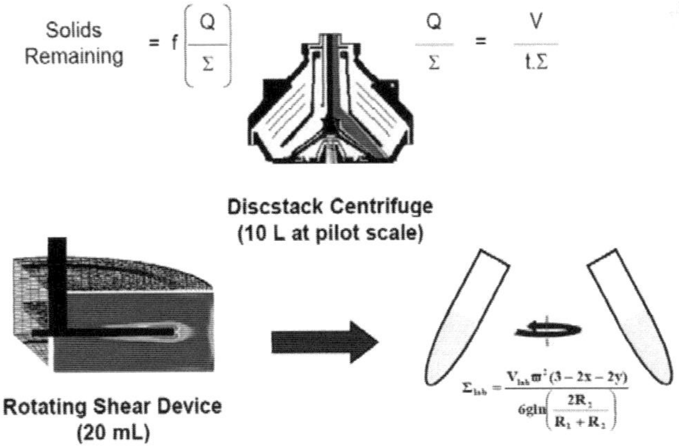

[Maybury et al (2000) *Biotechnol Bioeng*; Boychyn et al (2001) *Chem Eng Sci*]

Illustration of scale-down mimic in centrifuge operation

Industrial centrifuge feed zone

- Critical region causing feed material damage
- Not present in laboratory centrifuge
- Difficult to replicate as you reduce the size of the centrifuge

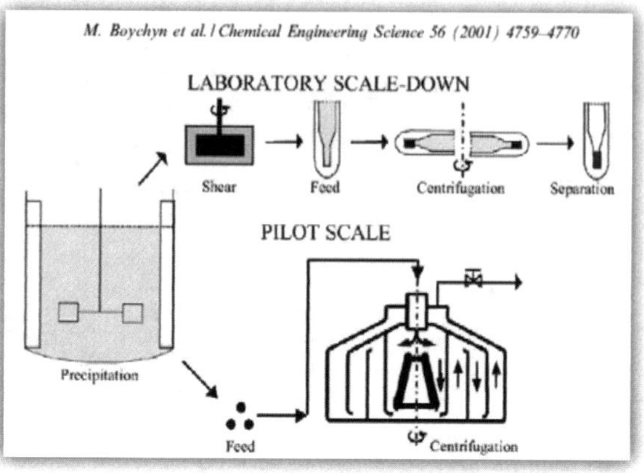

How will my material be affected by process shear?

USD sheared-centrifuge material

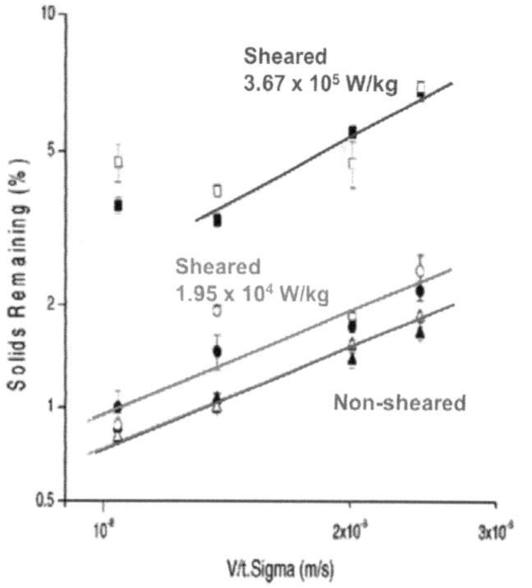

Hutchinson et al. Biotech. Bioeng., Vol. 95, No. 3, 2006

Predictions at larger scale

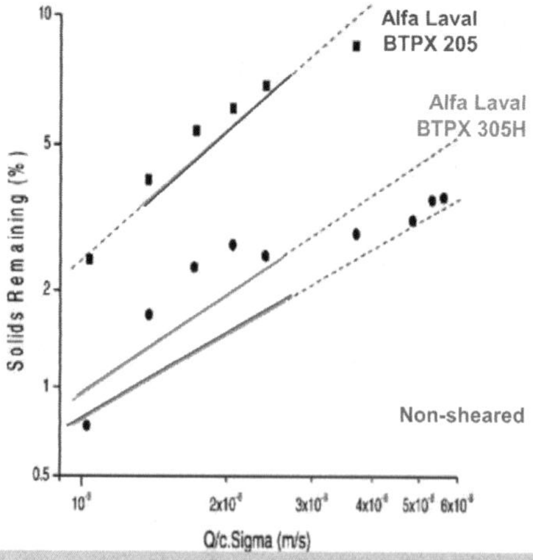

Laboratory experiments using non-sheared material over-estimate solids removal!

USD centrifugation:
shear + tubes/microplate + lab centrifuge

Material:
- Cell broth expressing therapeutic protein
- USD Shear Device
- Microwell plate

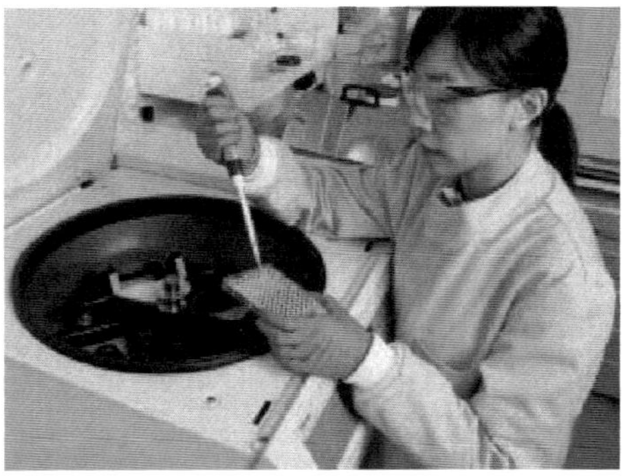

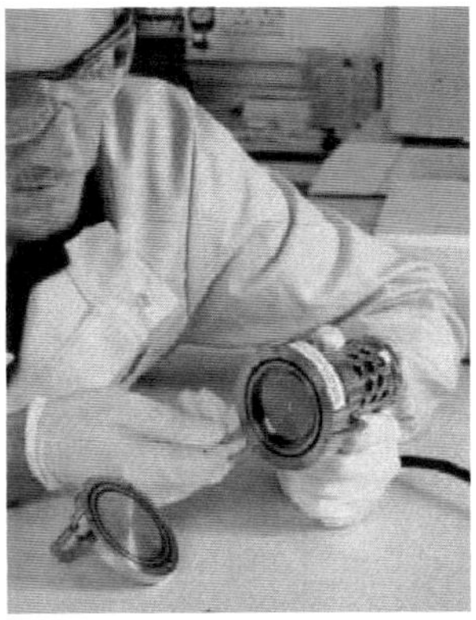

USD scale
- Sample volume: **20 mL**
- Shear exposure: low-high (equivalent to different types industrial centrifuges)
- Bench-top centrifugation: range of rcf and spin times
- Possible use of microplate in centrifuge

Case study 3: mAbs recovery on USD platform

USD + CoGs Analysis
- Compared low-sheared and high-sheared process materials
- Cover wide experimental space for unit ops in 1 day
- CoGs analysis for 200 kg and 2 tonne product

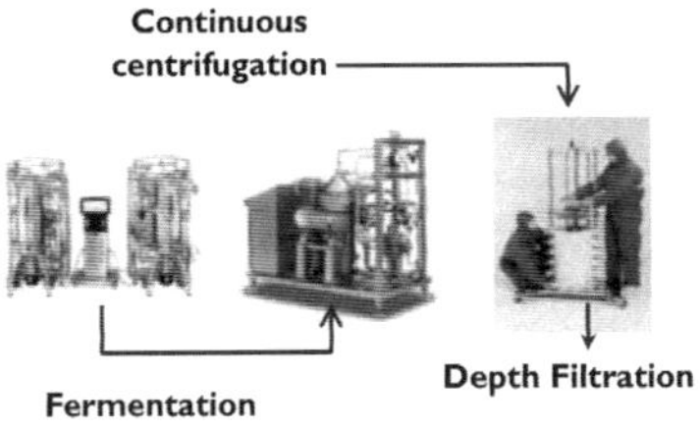

20 mL
→ 1 mL

USD Shear device
+ USD centrifugation

- ***Clarification***
- ***Titre***

1-4mL → ≤1 mL

USD Depth Filtration

- ***Vmax (L/m^2)***
- ***Titre***

Analysis and results

- mAbs typical process
  - Fixed equipment assets for both scenarios (low-, high-shear)
  - Includes capital cost depreciation (stainless steel), labour, media, buffer, filters, resin, etc
- 20% lower filter capacity for high-sheared material
  - Up to 20% higher solids remaining
- Higher titre for high-sheared material ???
  - Half-antibodies (Reid et al. 2010 *Biotech. Bioeng.*)
  - Loss of full antibody due to shear!
- 3-6% higher CoGs for high-sheared material

Cost Analysis

- Difference of $130K per 5-batch campaign (2 tonne plant)
- Opportunity cost (sales price versus loss of sales due to lower yield)!

Sales Price ($)	Cost of product/sales loss (at 200 kg plant; ~12kg less product due to high shear)	Cost of product/sales loss (at 2 tonne plant; ~131 kg less product due to high shear)
$ 500/g	$ 5.9 M/year	$ 66 M/year
$ 1000/g	$11.7 M/year	$ 131 M/year
$ 4000/g	$ 46.9 M/year	$ 524 M/year

- Opportunity cost! eg.Time to market = -1yr
  - $100-$800M (@200kg)
  - $1 B - $8 B (@2 tonne)

Ultra scale-down

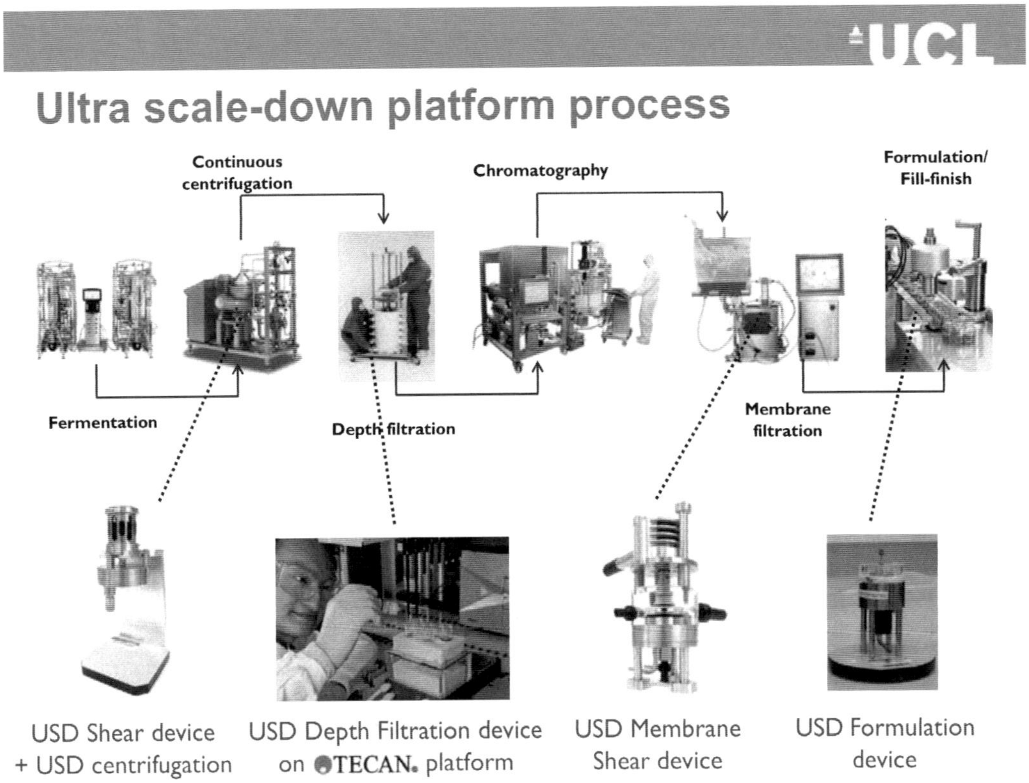
UCL
Ultra scale-down platform process
Continuous centrifugation
Chromatography
Formulation/ Fill-finish
Fermentation
Depth filtration
Membrane filtration
USD Shear device + USD centrifugation
USD Depth Filtration device on TECAN. platform
USD Membrane Shear device
USD Formulation device

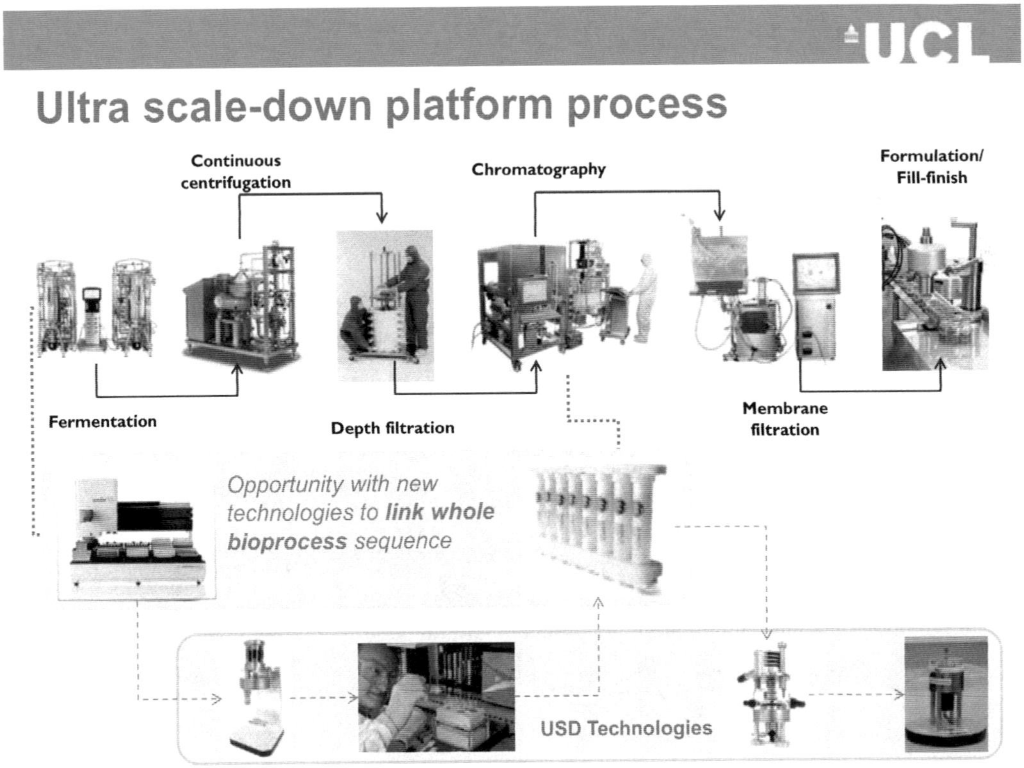
UCL
Ultra scale-down platform process
Continuous centrifugation
Chromatography
Formulation/ Fill-finish
Fermentation
Depth filtration
Membrane filtration
Opportunity with new technologies to **link whole bioprocess** sequence
USD Technologies

UCL

What is the value USD tech on users?

Feedback from industry users:

- Reduction of development time up to 50%
- 30% - 50% resource savings on development project
- 30% - 50% increased capacity per year
- 20% savings in COGS at £200,000 per annum
- Savings of $60k for a 5-batch campaign
- Savings of $280k/product transfer/site
- Not having to invest/pay for/use a large scale facility (and associated fixed costs)

- **Independent calculations show predicted values along these lines**

UCL

Value of success and early promise

Alliances with big up-front payments, 2013

Company	Country	Partner	Country	Up-front payments (US$m)
AstraZeneca	UK	FibroGen	US	350
AstraZeneca	UK	ModeRNA Therapeutics	US	240
Celgene	US	OncoMed Pharmaceuticals	US	177
AbbVie	US	Ablynx	Belgium	175
Celgene	US	MorphoSys	Germany	155
Biogen Idec	US	ISIS Pharmaceuticals	US	100
Celgene	US	Acetylon Pharmaceuticals	US	100
Roche	Switzerland	Chiasma	Israel	65
Baxter	US	Cell Therapeutics	US	60
Eisai	Japan	Arena Pharmaceuticals	US	60
Roche	Switzerland	Molecular Partners	Switzerland	59
Amgen	US	Les Laboratoires Servier	France	50
Daiichi Sankyo	Japan	Amplimmune	US	50

Source: EY, MedTRACK and company news.
Table includes only transactions for which up-front payments were publicly disclosed.

UCL

Towards an effective and *efficient* biopharmaceutical development

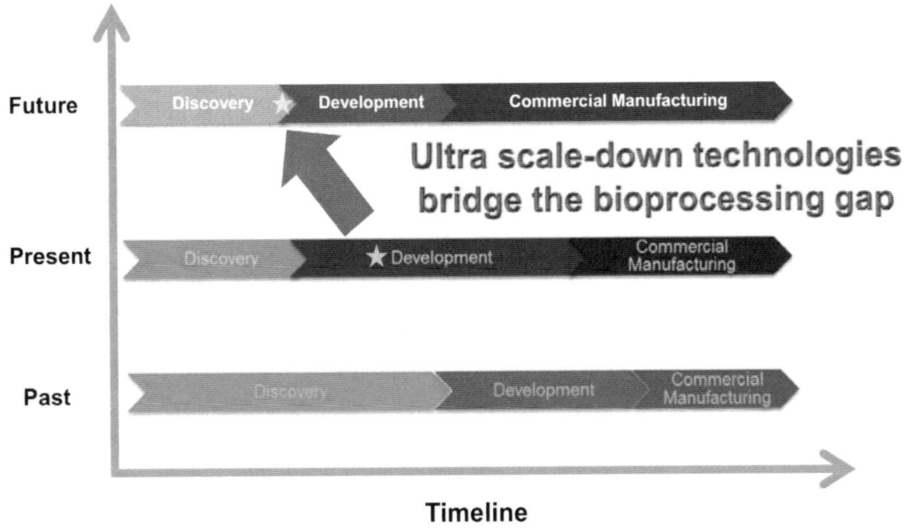

UCL

HIGHER EDUCATION hefce FUNDING COUNCIL FOR ENGLAND

Catalyst Fund

Enabling Biomanufacture across the Life Science Value Chain

Dr Andrea Rayat andrea.rayat@ucl.ac.uk
Dr Alex Chatel alex.chatel.09@ucl.ac.uk
Bioprocessing Enterprise Fellows

Prof Gary Lye g.lye@ucl.ac.uk
Prof Mike Hoare m.hoare@ucl.ac.uk
Advanced Centre for Biochemical Engineering

mbi modular training for the bioprocess industries

UCL ENGINEERING Change the world

The Queen's Anniversary Prizes 2013

Back-up slides

Bioprocessing Service Offerings for Industry

- **Bioprocess creation**
  - Ultra scale-down (USD) projects
    - USD technologies supply and support
    - USD-enabled bioprocess development
    - Consulting on USD applications
  - Single-use technologies
- **Bioprocess optimisation and scale-up**
- **Training and knowledge transfer**

Achieving Bioprocess Quality through Shared Expertise and Smart Technologies

Contact us for more information:

Dr Andrea Rayat & Dr Alex Chatel
Bioprocess Enterprise Fellows

usdtechnologies@ucl.ac.uk

USD supports your business case ...

- **Reduce cost** of development projects; enable early assessment of manufacturability
- **Increase capacity** to take on more development projects
- **Shorten development timeline,** and therefore allowing for timely project decisions (go, no-go)
- Reach the market earlier and maximise value of discoveries by longer patent protection post drug approval
- **Reduce cost of goods** at full-scale
- Check **material performance during industrial scale manufacturing/tech transfer/facility transfer**, avoiding downtime in production, by mimicking operation with USD

USD enables process development...

- Prepare material at small scale that is **representative of the full manufacturing scale**
- **Provide insights** into how the environment affect process material
- The ability to gain process understanding **earlier** than traditionally possible
- **De-risking process development** and ensuring **robust technology transfer** across scales.

Ultra scale-down applications

- Ultra scale-down studies of **effect of shear**
  - •Crystals
  - •Flocs
  - •Genetically modified plants
  - •Human cells for therapy
  - •Mammalian cells
  - •Precipitates
  - •Proteins and Enzymes
- Ultra scale-down studies of continuous centrifugation
  - •Algae
  - •Bacterial cells
  - •Flocs
  - •Mammalian cells
  - •Precipitates
- Ultra scale-down studies on depth filtration
  - •Bacterial cells
  - •Flocs
  - •Mammalian cells
- Ultra scale-down studies on membrane separation
  - •Algae
  - •Bacterial cells
  - •Human cells for therapy
  - •Protein solutions
  - •Viruses
- Ultra scale-down studies on formulation
  - •Freeze-drying formulation
  - •Protein solutions

Microwell Systems for High Throughput Bioprocess Design (HTBD)

Early systems | Characterisation/ Scale Translation | Scale-up demonstration

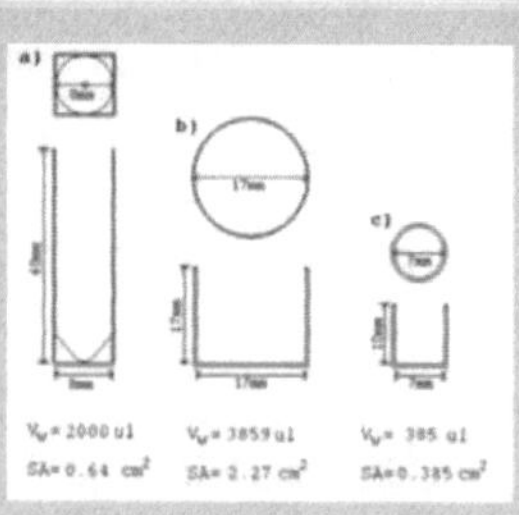

[Lye *et al* (2003) *Trends Biotechnol.*; Micheletti & Lye (2006) *Current Opin. Biotechnol.*]

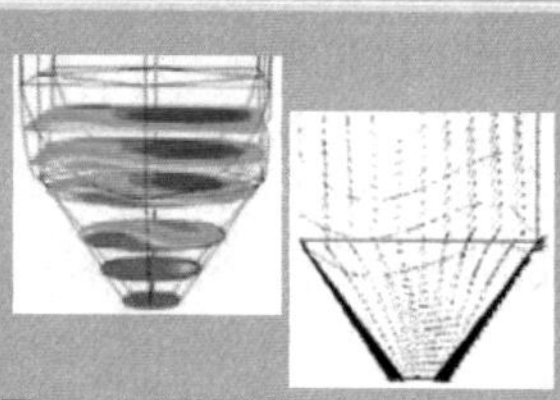

[Zhang *et al* (2008) *Biochem. Engng. J.*]

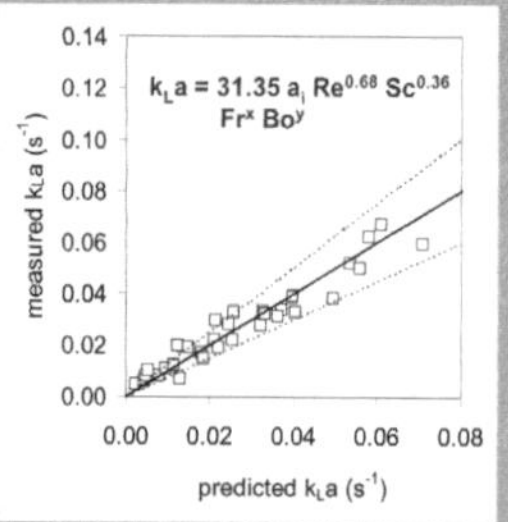

[Doig *et al* (2005) *Chem. Eng. Sci.*]

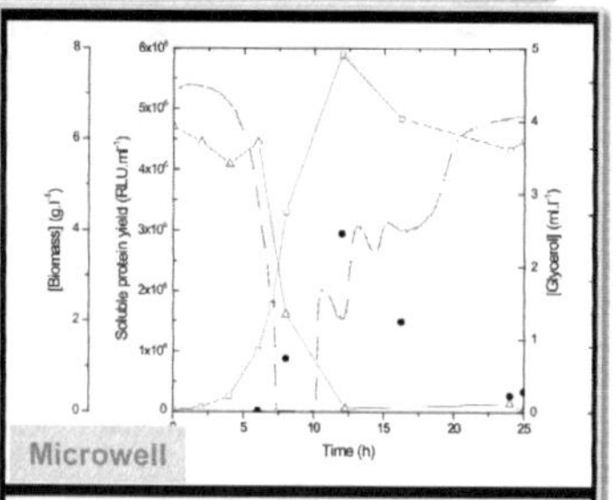

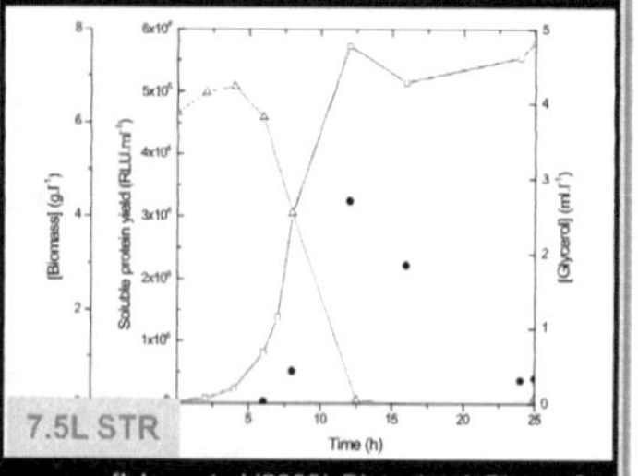

[Islam *et al* (2008) *Biotechnol. Bioeng.*]

The Same Considerations Apply to DSP Operations e.g. Microscale Chromatography

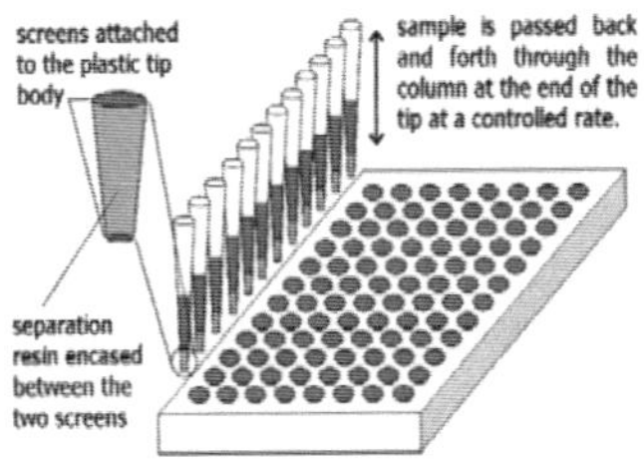

PhyNexus Tips

Wenger *et al* (2007) *Biotechnol. Appl. Biochem.*
Wenger *et al* (2008) *Biotechnol. Progr.*

Robo Columns

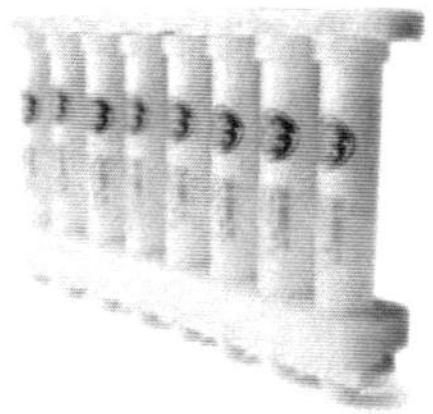

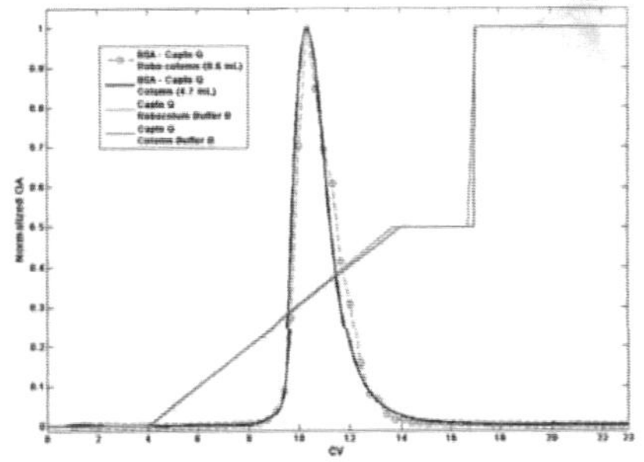

Konstantinidis *et al.* (2014) *HTPD Olympiad*

USD membrane separation – how will my product interact with membranes?

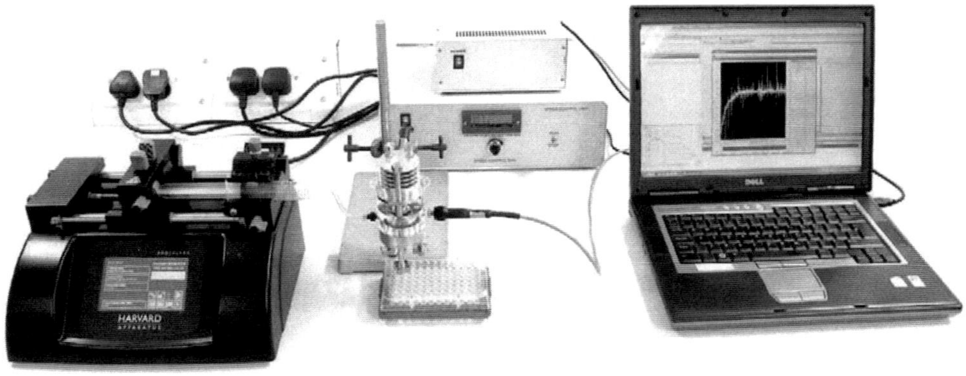

USD membrane separation – how will my product interact with membranes?

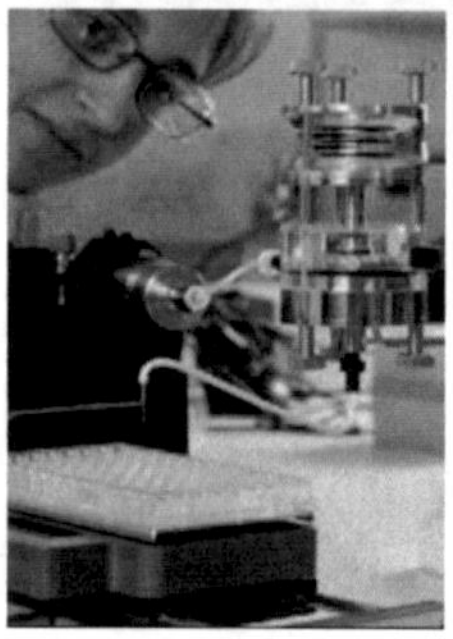

Material:

- Human neonatal foreskin fibroblast cell line
- USD Membrane Shear Device
- Buffer exchange/formulation media of choice

USD scale

- Sample volume: **1.5 mL**
- Variables: flux (flow-rate), shear rate

Result: insight into effect of processing on cell morphology

Pre-processing **Post-processing**

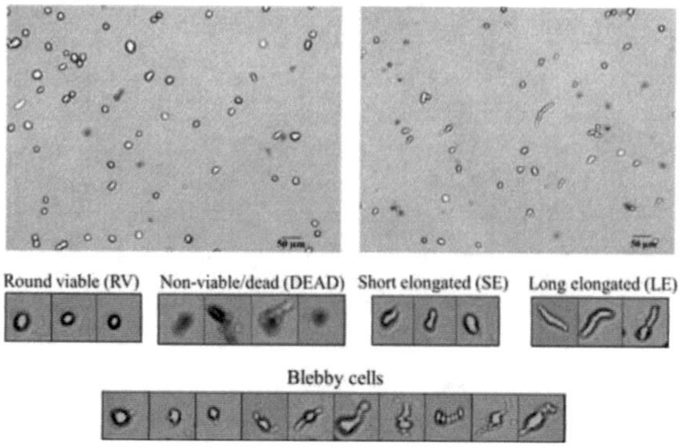

Masri MF, Lawrence K, Jaccard N, Wall I, Hoare M (2013) Ultra scale-down approach to membrane separation procedure of human cells for therapy as a predictive tool for scale-up IChemE BESIG (First prize)

USD centrifugation – how will my material be affected continuous centrifugation?

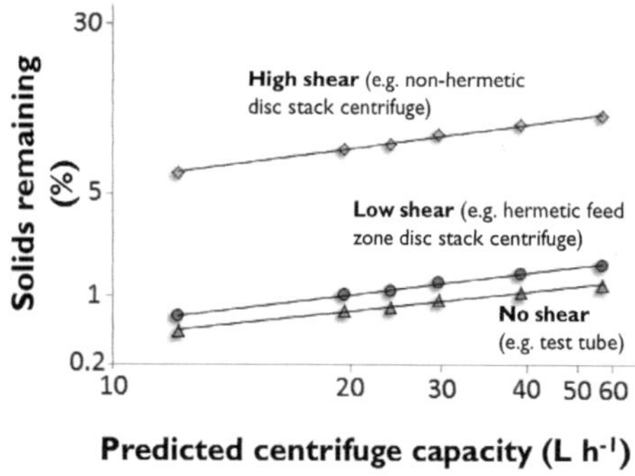

Tait, A S, Aucamp,JP, Bugeon,A, Hoare,M. (2009). Ultra scale-down prediction using microwell technology of the industrial scale clarification characteristics by centrifugation of mammalian cell broths Biotech Bioeng 104(2): 321-331. .

Result: successful prediction of process scale clarification

Rapid Process Development Enabled By Automated, Single Use High Throughput Technologies
Kistler, C.,
Originally presented on: 3/3/2015 1:15:00 - 1:40:00

Rapid Process Development Enabled by Automated, Single Use High Throughput Technologies

Christopher Kistler

Bioprocess Technology and Expression, Bioprocess Development, Merck and Co. Kenilworth NJ

Efforts continue to speed the progress of new biologics from discovery through preclinical development to product launch. This is challenged by the move to novel modalities and managing the existing pipeline of products with limited resources. To provide rapid and right first time development, Merck Bioprocess Development applied automation and singe use technologies from end to end, covering from cell line development to purified drug substance. For clone screening, laborious shake flask studies were replaced by a spin tube system with automated sampling and feeding capability. Further, inclusion of the Ambr15™ in the clone evaluation stage increased capacity and decreased process time both by 2X. Early stage upstream process development timelines were decreased 3-6X through the implementation of the Ambr15™ for medium and feed screening then to the Ambr250™ for process development. In addition, single use technology was implemented up to 2000L scale to accelerate the supply of drug candidates for development. Downstream processing development was moved from a platform that included small and preparative columns to mini column and micro tip formats operated on liquid handling arrays. This enabled execution of high resolution experiments to probe the design space, which resulted in more data with fewer resource investments in an equivalent time period. These technologies were supported by the development of high throughput analytical assays and tools. Titer and host cell protein assays were moved from an ELISA/HPLC platform to a rapid microfluidic Gyrolab™ platform. This change alone reduced assay time 3X while increasing throughput 10X. Taken together, the investment in automated high throughput technologies well positioned Merck Bioprocess development to execute on its mission of delivering robust and well characterized processes with accelerated timelines.

Improving Manufacturing Network Productivity: Overcoming the Downstream Bottlenecks

Conley, L., Biogen

Originally presented on: 3/3/2015 1:40:00 - 2:05:00

With a maturing product pipeline and higher dose (and thereby production) requirements for many protein therapeutics, internal manufacturing capacity will be exceeded in the coming years, requiring additional manufacturing facilities and/or outsourcing production to CMOs. Improved manufacturing processes are being developed using a combination of technologies established in other industries, such as continuous processing, and newer concepts to improve manufacturing network productivity and enable process fit into existing plants. In addition, advancements in cell line and culture technologies have led to higher titers (>5 g/L) for monoclonal antibodies coupled with shorter bioreactor duration. This has necessitated the development of higher throughput downstream processes to prevent purification from being the throughput bottleneck in the manufacturing process. The primary bottlenecks reside in the initial purification (capture) step using protein A affinity chromatography with relatively low antibody binding capacity, and in the intermediate product pools and buffer volumes exceeding the capacity of existing production vessels throughout the manufacturing process. Several technologies were evaluated to address these bottlenecks and help improve productivity including (i) newer generation protein A resins with higher binding capacity (ii) semi-continuous and continous modes of chromatography for antibody capture on protein A and (iii) single pass tangential flow filtration coupled to purification steps to enable in-line concentration of intermediate product streams. This presentation will show how integration of these process improvements into the purification process can facilitate debottlenecking and improve facility fit. Ultimately, these process improvements are expected to extend the lifespan of current manufacturing capacity, minimizing the need to retrofit existing facilities or to build additional manufacturing facilities in the future.

Beyond Numerical Tools: Control Strategy Principles that Assure Product Quality Throughout Development

Ruesch, M. N.,

Originally presented on: 3/3/2015 0:00:00 - 0:25:00

At Pfizer we have taken a lifecycle approach toward implementing control strategies for biotherapeutic products per ICH Q10. Developing process and product understanding to implement a holistic control strategy is critical to ensure consistent, quality products. The challenges of implementing control strategy principles for diverse modalities in early and late development are quite different. In early phase, targeted application of enhanced product characterization tools can influence molecular design and are critical to inform process development. Application of these tools needs to be focused and informed by platform knowledge, and ideally the information is rapidly available due to the short timelines and resource constraints in early development. By later stage, extensive product and process understanding needs to be translated into an understandable, robust, and defendable control strategy. Case studies where enhanced analytical characterization and deeper quality attribute understanding have been applied to phase appropriate control strategies will be discussed.

Efforts to simplify the sometimes complex quality attribute and control strategy tools are critical to ensure we are focusing on the scientific understanding of the product quality attribute profile for complex molecules. Looking forward, it is particularly critical to continue technical and organizational efforts to move aspects of control and product understanding closer to the point of impact during all stages of process development, rather than traditional reliance on retrospective and final release testing. Pfizer's efforts in these areas will also be presented.

Unlocking the Power of Analytics in Support of QbD in Development

Bak, H.,

Originally presented on: 3/3/2015 0:25:00 - 0:50:00

In preparation for commercialization, Regeneron typically introduces a new high titer cell line and optimized production strategy for improved plant capacity and cost-of-goods while meeting or exceeding quality from first in human manufacturing process. For this work, the principles of Quality by Design is leveraged for targeted, rational development. The presentation will provide examples of how analytical support is applied to serve a persistent focus on critical quality attributes to drive this development effort.

Understand & Control Product Quality Attributes in Cell Culture
Ouyang, A., Eli Lilly & Company
Originally presented on: 3/3/2015 0:50:00 - 1:15:00

Understand & Control Product Quality Attributes In Cell Culture

Anli Ouyang, Bioproduct R&D, Eli Lilly and Company

In preparation for commercialization of therapeutic protein manufacturing, cell culture process is developed and characterized to ensure consistent delivery of high quality protein. Drug substance and drug product quality attributes can be impacted by the cell culture process. Two case studies will be presented to demonstrate how Lilly developed new tools to understand and control critical quality attributes in cell culture.

In a fusion protein program, a protein cleavage form was considered as critical quality attribute due to its lack of potency. This cleavage form cannot be removed by downstream purification process and presented at relatively large percentage (10%) in drug substance. It is important to develop a robust cell culture process so that this product related impurity can be minimized and controlled. Traditional tools (i.e. protease inhibitors, pH and temperature etc) had limited impact on this cleavage. However, we found that osmolality and pCO_2 controls were important to reduce this impurity in drug substance. With implementation of the new control strategy, the cleavage level reduced by 50% and was successfully scaled-up.

For the second case study, , we will share our observations that cell culture process changes had a significant impact on high protein concentration (> 100 mg/ml) drug products stability profiles in several molecules. HMW aggregates growth rate in drug product increased significantly after the cell culture medium change though aggregates level in drug substance were similar. Lilly developed a surrogate assay for drug substance as an indicator to aggregates on drug product stability. By using this surrogate assay, a large number of cell culture factors were screened and a few levers (i.e. media components and process conditions) in cell culture were identified to impact drug product aggregates growth, which were later confirmed in real time stability studies.

Closing the gap- Addressing similarity challenges of a biotherapeutic with the innovator product

Seth, G.,

Originally presented on: 3/3/2015 1:15:00 - 1:40:00

Biosimilar development holds the promise of increasing access of low cost medicines to patients in emerging markets and the world at large. At the core of a biosimilar program is the ability to ensure that the physico-chemical, pre-clinical and clinical data generated for the biotherapeutic is similar to the innovator product with respect to the structural properties, safety and efficacy. In this presentation, taking an in-house case study, a target-directed strategy for development of a biosimilar therapeutic will be outlined. Key aspects of process improvements and impact of both upstream and downstream process parameters on final product quantity and quality will be discussed. Process optimization efforts presented were geared towards minimizing process development time in a time sensitive biosimilar development space, whilst meeting regulatory standards for acceptable product profile of the biosimilar. Strategies discussed during this presentation will help accelerate biosimilar development in both emerging and regulated markets.

A Chinese Hamster Ovary Cell Host Cell Protein That Impacts PS-80 Degradation

Lee, K. H., University of Delaware
Lenhoff, A. M., University of Delaware
Levy, N., University of Delaware
Valente, K. N., University of Delaware
Originally presented on: 3/3/2015 1:40:00 - 2:05:00

The recent sequencing of the Chinese hamster and Chinese hamster ovary (CHO) cell genomes enables cell engineering strategies to address a wide variety of problems encountered in biopharmaceutical manufacturing. One particular application involves studies of CHO host cell proteins (HCPs) that may be difficult to remove for a variety of reasons. The presence of HCPs is regulated for patient safety concerns but may also have an impact on product quality in the context of formulation. Polysorbates are a class of non-ionic surfactants that are added to biopharmaceutical formulations to improve the stability of therapeutic proteins by limiting aggregation and surface adsorption. Monoclonal antibody formulations often incorporate polysorbate 80 (PS-80) to prolong the shelf-life of drug products. PS-80 degradation over time can impact the stability of those drug products.

We studied a range of CHO HCPs that are difficult to remove for a variety of reasons including HCPs that coelute with various monoclonal antibodies, HCPs that may bind to various monoclonal antibodies, and HCPs that change in expression as a function of cell age. We identified one HCP, HCP A, that is particularly difficult to remove. Based on the function of this protein, we hypothesized that the presence of this molecule in PS-80 containing drug formulations may result in increased degradation rates of PS-80. We used an interfering RNA approach to reduce the expression of HCP A in CHO cells. Those modified cells exhibited a significantly lower level of expression of HCP A than control cells, as measured by a mass spectrometry-based assay that was developed to quantify HCP A expression, without significantly impacting cell viability or cell density. The HCP from cells modified with the interfering RNA exhibited substantially reduced hydrolysis of PS-80 relative to HCP derived from control cells. Additional observations based on genome editing strategies to knock out the HCP A gene will be discussed.

Continuous Processing for Pre-Clinical and Clinical Manufacturing: Balancing Speed, Cost, and Quality

Coffman, J., Boehringer Ingelheim
Lin, H., Boehringer Ingelheim
Wang, S., Boehringer Ingelheim
Godfrey, S., Boehringer Ingelheim
Orozco, R., Boehringer Ingelheim
Yildirim, S., Boehringer Ingelheim
Setiabuti, G., Boehringer Ingelheim
Vogel, J., Boehringer Ingelheim
Originally presented on: 3/4/2015 0:50:00 - 1:15:00

Continuous processing has been touted as a new paradigm in commercial manufacturing of biologics. Significant effort has been spent on achieving optimal commercial processes. These efforts have resulted in processes with ten-fold increases in productivity and lower costs than typical batch processes.

Early stage clinical manufacturing, however, has different needs than commercial manufacturing. Cost and productivity, in and of themselves, are not as important as the ability to manufacture multiple candidates quickly and robustly, yet with the assurance of scaleability for commercialization.

A launch capable, high productivity mammalian perfusion process appears to be easier to develop than fed batch systems with equivalent productivity. This facile development of perfusion bioreactors and a robust, effectively continuous platform downstream can allow the inexpensive production of regulatory toxicology material prior to candidate selection for multiple candidates, shortening the time to the IND by five to eight months. The product quality is acceptable for new biological entities. The productivity of such a perfusion process will be demonstrated with the production of 1kg of material in a 100L SUB in 15 days. The challenges associated with using continuous processing in early stage clinical supply will be discussed.

Developing a Leading Open-Access Integrated Biologics Development and Manufacturing Platform to Expedite Globe Development

Weichang Zhou, Ph.D., Vice President, Biologics Process Development
WuXi AppTec, Shanghai, China

Accelerating Biopharmaceutical Development
Scottsdale, Arizona, USA, March 4, 2015

药明康德 WuXi AppTec

Presentation Outline

药明康德 WuXi AppTec

- **Overview of WuXi AppTec and WuXi Biologics**

- **Integrated biologics development and manufacturing**

  #1. Establishing leading technology platforms: fed-batch, perfusion (continuous processing) and concentrated fed-batch for diversified product modalities
  #2. Building and operating manufacturing facilities using single-use technologies
  #3. Technology transfer and manufacturing of a novel Mab
  #4. Development and manufacturing of a novel ADC

- **Concluding remarks and acknowledgements**

Confidential Information

OUR JOURNEY

2000: Founding of WuXi

VISION

Building an open-access capability and technology platform to enable anyone and any company to discover and develop healthcare products to benefit patients

2001

4 founders, 7,000 sq ft lab space

WuXi's first lab, China's first modern chemistry hood

OUR JOURNEY

2014: Global Platform & Comprehensive R&D Enabling Capabilities

2014

9,000 employees

5 million ft^2 existing & upcoming R&D space

leading capability and technology platform

Small Molecules

Biologics

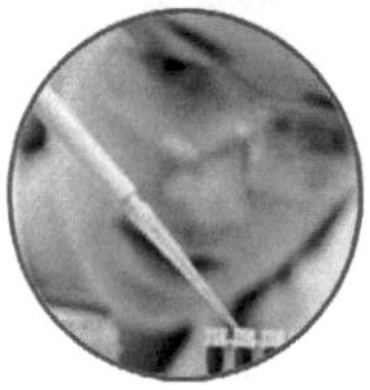

Cell Therapies

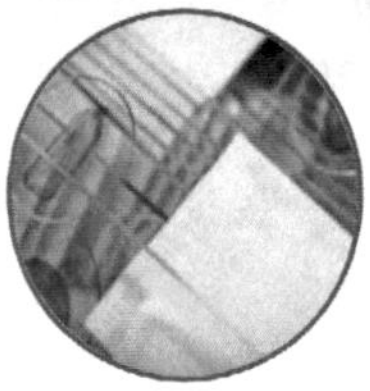

Medical Devices

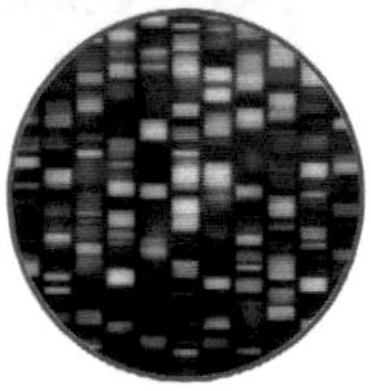

Genomics

WuXi's Growth Through Innovation

14 Years of Diligent Capability Building

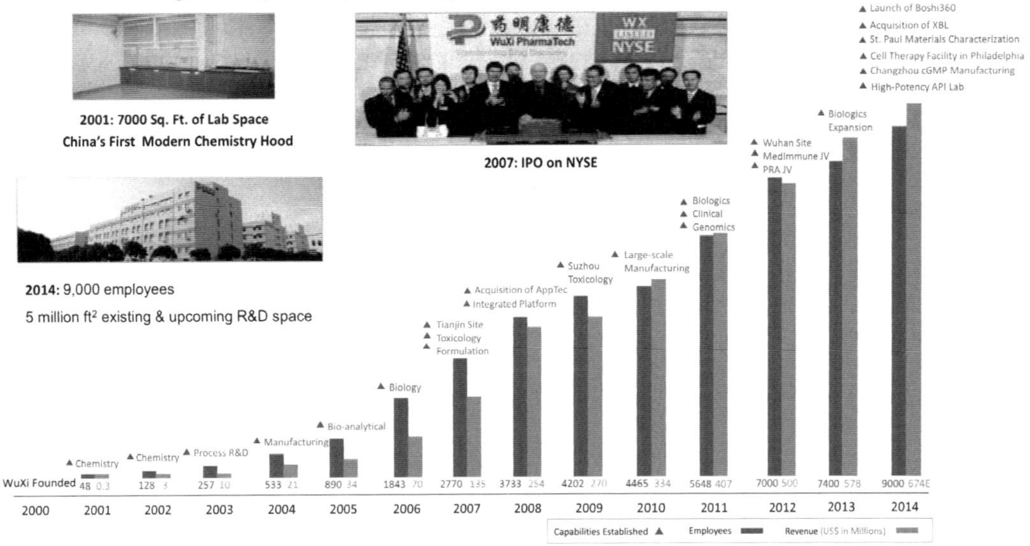

OUR PLATFORM

21 R&D Sites and Offices Worldwide

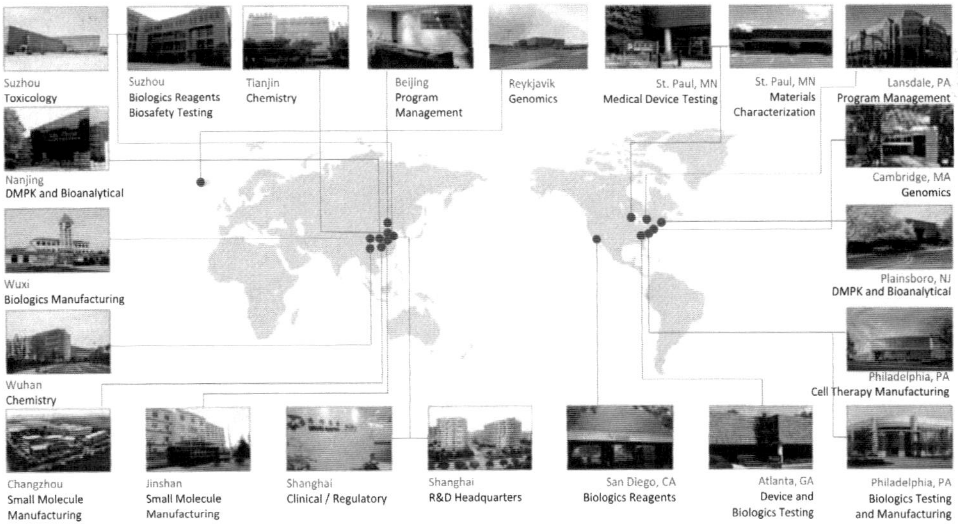

OUR PLATFORM

Biological Drug Enabling Platform

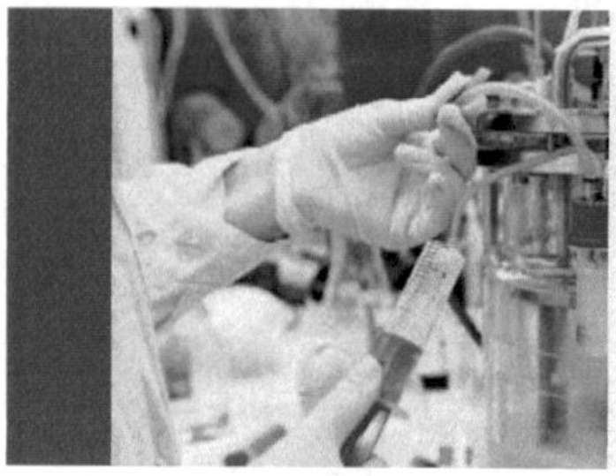

- The first cGMP biologics manufacturing facility in China that is compliant with global regulatory standards
- > 80 global customers, including 7 of the 20 largest pharma companies
- The first biologics drug manufactured in China for client's U.S. late stage clinical trials
- Contribution to 2 BLA filings for commercial launch and 35 IND filings for human clinical trials by December 2015

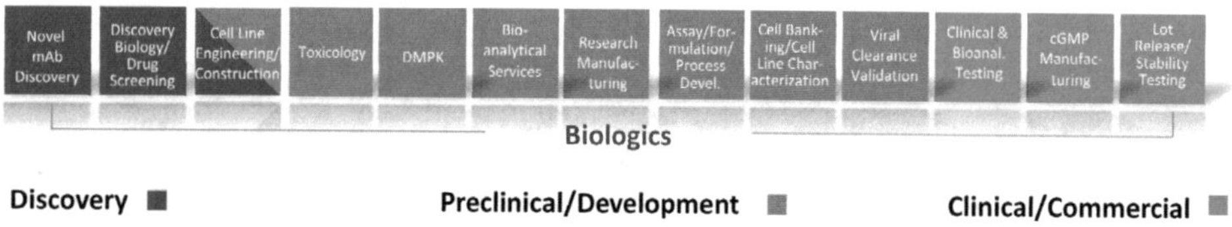

WuXi Biologics (Shanghai and Wuxi, China)

- **Highly capable team**
  - 850+ people, including about 100+ returnees with US/EU industry experience
  - World's largest CDMO development team with a team of 500 people
- **Rapidly expanding client base**
  - > 80 clients including 7 of the top 20 large pharma companies
  - Clients from US, EU, Israel, Japan, Korea, Singapore, Australia, India and China
- **Extensive project portfolio**
  - Product types include mabs, Fc fusion proteins, ADCs and glycosylated proteins
  - Currently working on 12 mAb discovery projects, 40 mAb/rprotein development projects from cell line dev to IND filing or BLA filing
- **Expanding services & platforms**
  - Extensive integrated ADC capabilities, 100+ g scale cGMP conjugation completed
  - Large-scale ATF-based perfusion program (6 on-going perfusion projects)

Confidential Information

Topics 1: Establishing Leading Technology Platforms

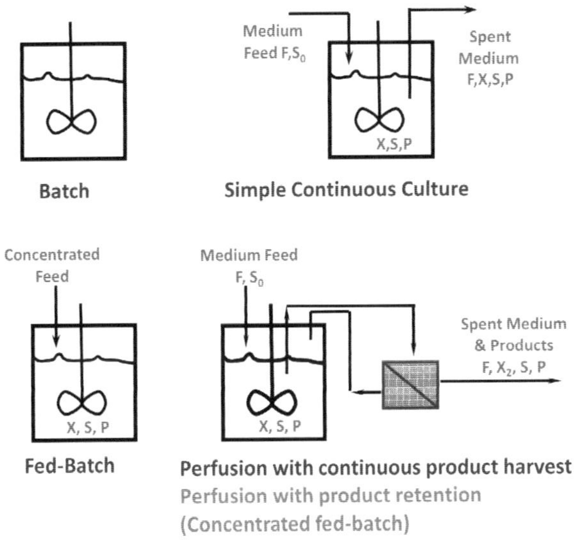

- **Fed-batch: stable products**
- **Perfusion: liable and stable products**
- **Perfusion with product retention (concentrated fed-batch): stable products**

Establishing High Cell Density Perfusion Culture Technology Platform

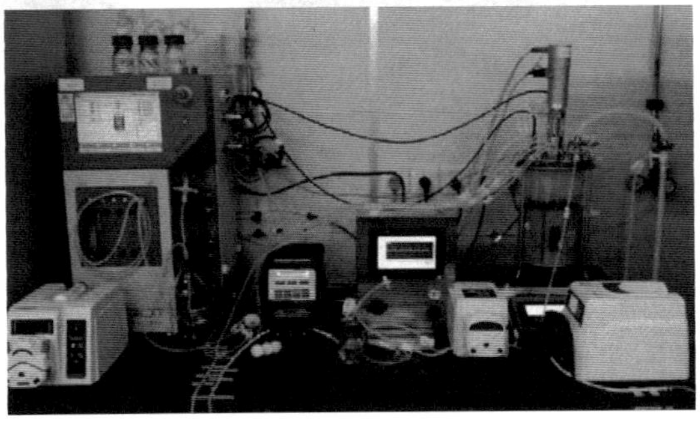

Lab scale ATF2 setup

- ATF-based perfusion culture platform
  Lab scale: ATF2 and ATF4
  Production scale: ATF6 & ATF10
  Steady state culture @ >40x10^6 cells/mL for up to >100 days
- Develop platforms for both labile recombinant proteins and stable Mabs
- Large-scale cGMP manufacturing at 125 or 1,000L scale

High Cell Density Steady State Perfusion Culture

药明康德 WuXi AppTec

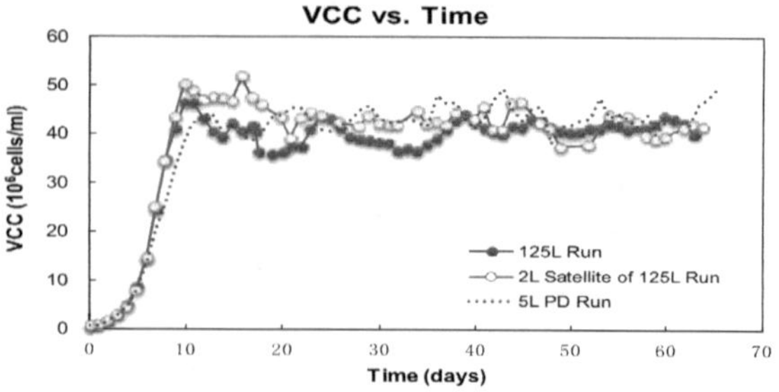

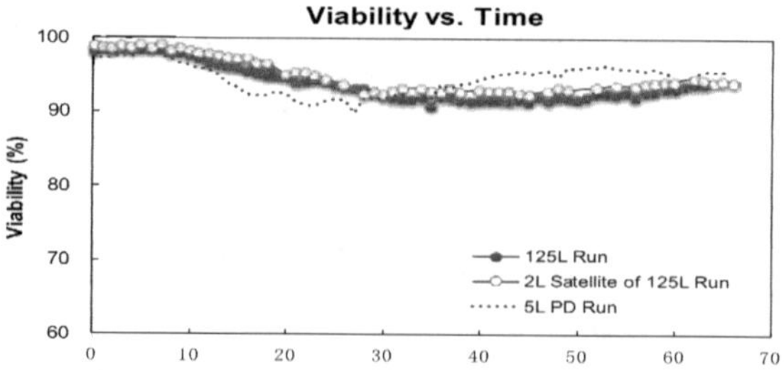

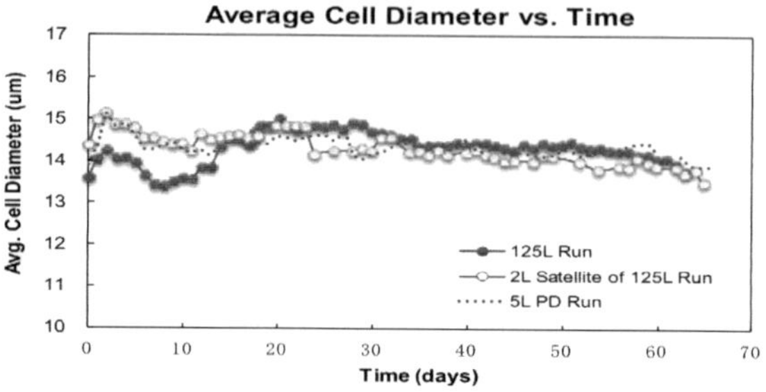

- Scale: 125L in 250L SUB, coupled with ATF6
- Steady state culture > 40×10^6 cells/mL for > 60 day
- Very high viability

Confidential Information

Steady State Perfusion Culture: Cell Metabolism

药明康德 WuXi AppTec

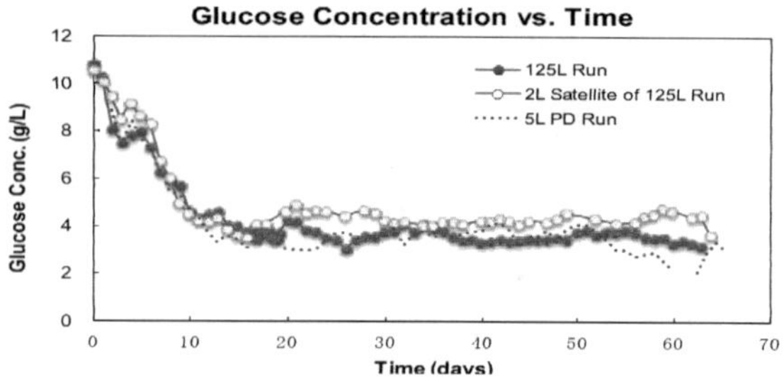

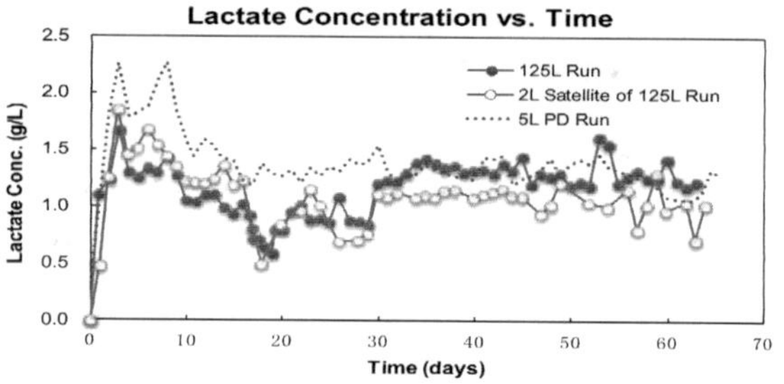

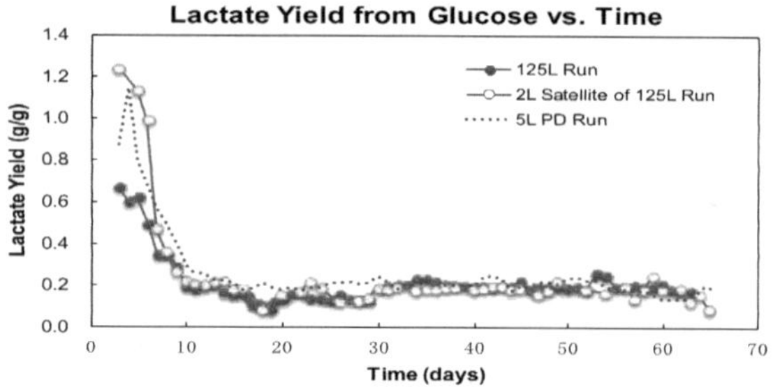

Developing Concentrated Fed-Batch Platform

- Background: initial fed-batch process is ~2 g/L for a Mab
- Direct adoption to concentrated fed-batch
- Similar culture duration as fed-batch, ~2 weeks (or longer)
- ATF-based cell and product retention
  - ATF filter pore size: 50KD
  - Continuous feeding with fresh basal and feed media (for certain duration of culture)

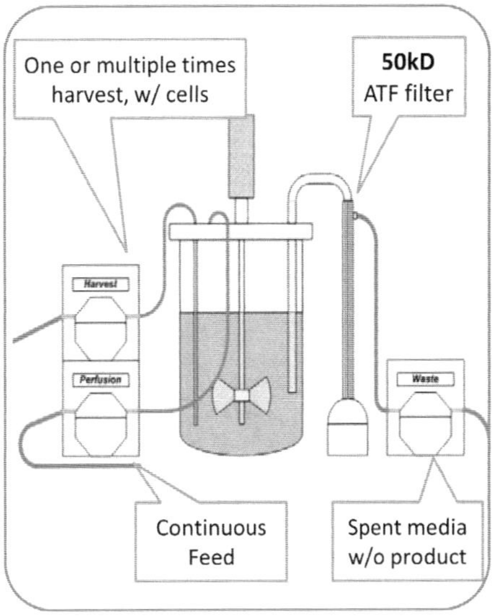

Comparison of Concentrated Fed-batch vs. Fed-batch Cell Growth

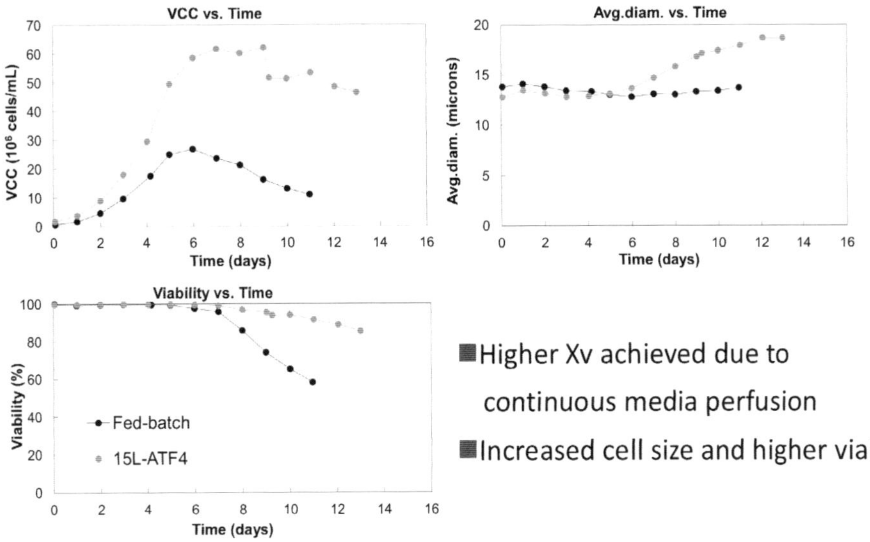

- Higher Xv achieved due to continuous media perfusion
- Increased cell size and higher viability

Comparison of Concentrated Fed-batch vs. Fed-batch Antibody Production

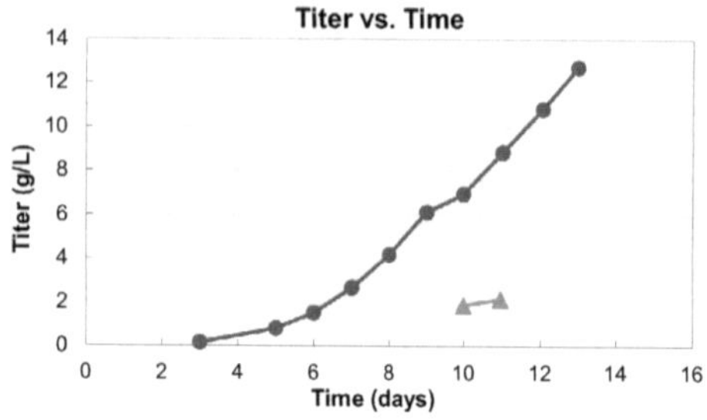

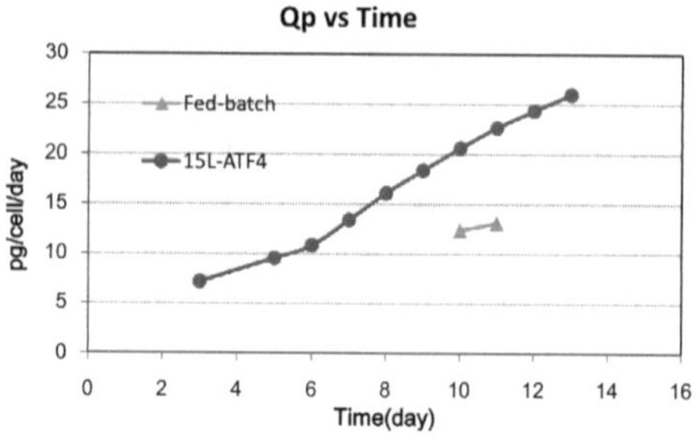

- 12.71 g/L vs. 2.0 g/L (less than 2 weeks)
- Qp increased throughout the culture for ATF run and was > 2x higher than fed-batch process

Summary I

- **High cell density perfusion culture platform established and scaled up with sustained high cell density and productivity**
- **Integration of continuous perfusion cell culture and product capture lays foundation for continuous bioprocessing**
- **Concentrated fed-batch culture takes advantage of perfusion in a fed-batch setting to achieve higher productivity**

	Quality	Speed	Cost
Fed-Batch	+	+++	+
Perfusion	+++	+	+++
Concentrated Fed-Batch	++	++	++

Topic 2: Building and Operating 1st cGMP Biologics DS & DP Manufacturing Plant with Single-Use Technologies in China

药明康德 WuXi AppTec

- Designed and reviewed by 4 large pharma
- Operational since Q12012
- First cGMP DS/DP lot shipped to US Q22014, and first batch of cGMP DS will be shipped to EU and Austria in 2015
- ISPE Facility of the Year: Honorable Mention 2014

Suite 1: 1x50L/200L/1000L all from Sartorius
Suite 2: 1x50L/200L/500L all from Sartorius
Suite 3: 2x2000L from Hyclone
Harvest from either bioreactor, single purification suite

Building Commercial Manufacturing Facility: Operational in 2016

药明康德 WuXi AppTec

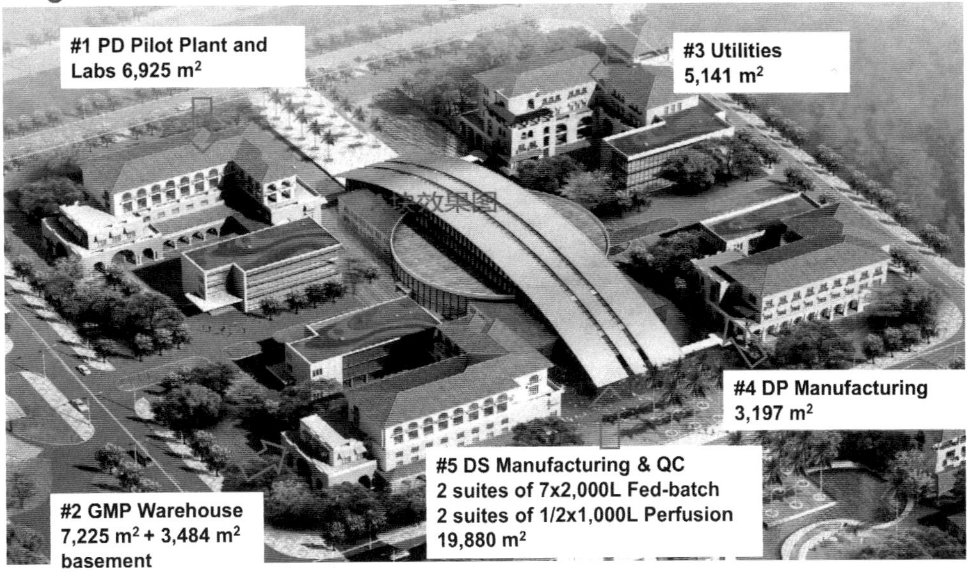

Topic 3: Technology Transfer and Manufacturing of a Novel Mab
WuXi-TaiMed Collaboration to Enable Ibalizumab Development

WuXi PharmaTech to Manufacture Ibalizumab for TaiMed Biologics

SHANGHAI and TAIPEI, August 29, 2012 /PRNewswire-Asia/ -- WuXi PharmaTech (NYSE: WX) today announced that it has a contract manufacturing agreement with TaiMed Biologics Inc. for the manufacture of Ibalizumab (TMB-355) in support of **Phase 2 and Phase 3 clinical trials globally**. Ibalizumab is a humanized monoclonal antibody and a member of an emerging class of HIV therapies known as viral-entry inhibitors. This work will be completed at WuXi's biologics manufacturing facilities in China and its biologics testing facilities in the United States.

- Cell line and process developed in the US, and scaled up to 2,000L scale

  Multiple GMP batches made for Phase I/II clinical trials in stainless steel bioreactors

- To produce GMP materials for PhII/PhIII global clinical trials in Wuxi's new GMP facility, using 2000L disposable bioreactors and the same purification process

- To support pre-BLA process characterization & validation work, initial product launch

Confidential Information

Tech Transfer Approaches/Strategies

- Comparability protocol governed the whole process transfer, optimization and manufacturing

  Established evidence that the manufacturing process and materials produced after transfer sufficiently comparable to the process and clinical materials produced and used in trials before the transfer

- Cell culture performance at 2,000L SUB, is the main technical consideration:
  Productivity and product quality comparability

- Two directions for 1st six months
  Scale down model transfer from client using bench-top glass bioreactors
  Disposable bioreactor evaluation

- 50L, 250L SUB to be evaluated before 2,000L SUB

Confidential Information

Successful Technology Transfer and Manufacturing Performance at 2000L SUB – Growth & Titer

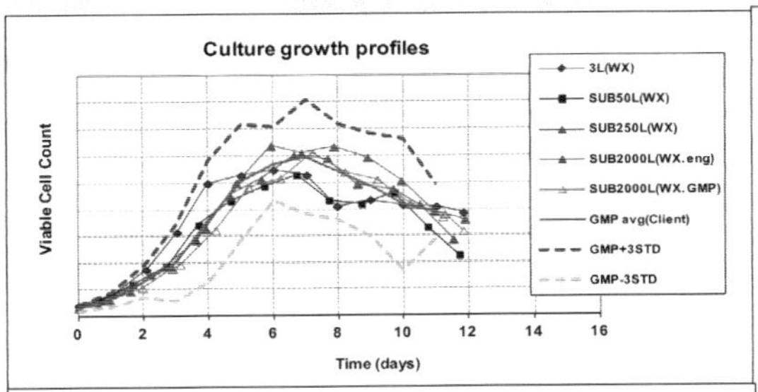

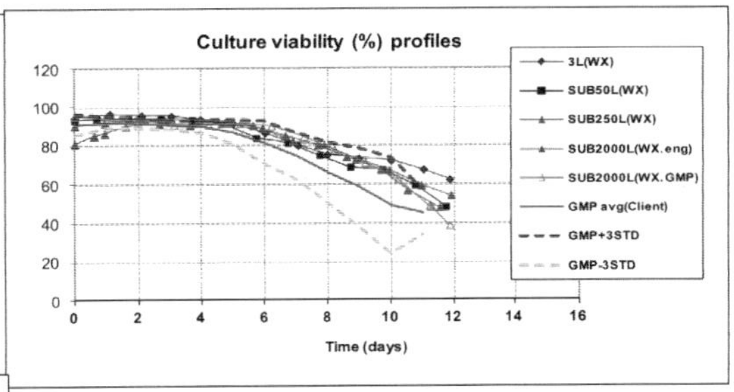

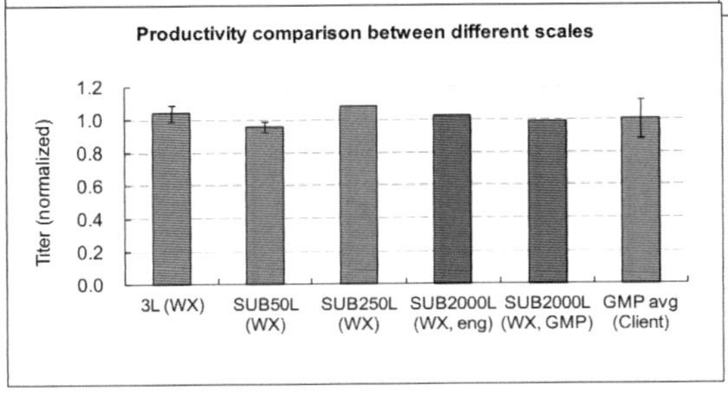

- Growth & productivity at 50L, 250L and 2,000L SUB (eng run & GMP run) matched very well with historical GMP data

Product Quality Comparability Assessment

- Comparability protocol submitted to FDA and accepted before 2,000L production
  To compare against historical stainless steel bioreactor results
- Initial comparability assessment done using 50L SUB and 250L SUB drug substance
  Process performance
  Extensive analytical comparison: release and characterization assays
  Real-time and accelerated stability comparison also part of comparability
- 2,000L engineering run DS also thoroughly analyzed
- DS & DP specification proposed based on historical GMP run data and WuXi data

Breaking News of 1st batch of DS and DP of Ibalizumab Manufactured at WuXi used in US Trials

药明康德 WuXi AppTec

FDA Approves Ibalizumab Manufactured by WuXi PharmaTech for Treatment of Patients on Expanded Access

SHANGHAI and TAIPEI, May 5, 2014 -- WuXi PharmaTech (NYSE: WX) and TaiMed Biologics today announced that the FDA has approved the first batch of the ibalizumab (TMB-355) drug substance and sterile drug product manufactured at WuXi's biologics facilities for ongoing treatment of patients under investigator-sponsored IND's. This is the *first reported FDA approval of a sterile biologics product manufactured in China for use under a U.S. IND, which marks an important milestone of WuXi's investment to expedite global development of innovative biologics.*

Ibalizumab is a humanized monoclonal antibody and a member of an emerging class of HIV therapies known as viral-entry inhibitors. It is being developed by TaiMed Biologics for the treatment of HIV/AIDS infection.

The drug substance of ibalizumab was manufactured at WuXi's cell culture manufacturing facility, which recently received an honorable mention facility-of-the-year award by ISPE. The ibalizumab drug product batch approved by the FDA was manufactured at the newly completed state-of-the-art cGMP parenteral manufacturing facility located within the same campus as the cell culture facility at Wuxi city. The automated filling line can accommodate 2-50mL liquid and lyophilized vial products for global clinical trials and product launch.

Confidential Information

Breaking News of 1st batch of DS and DP of Ibalizumab Manufactured at WuXi used in US Trials

药明康德 WuXi AppTec

BIOWORLD™ TODAY
BREAKING NEWS BULLETIN

Milestone: FDA approves first ever China-made biologic for use in U.S. trials

By Cornelia Zou, Staff Writer

HONG KONG - The U.S. FDA has, for the first time ever, approved the use of a biologic product made by a Chinese company in U.S. investigational new dru

R&D services giant Wuxi Pharmatech Inc. (NYSE:WX) got the green light for HIV therapy drug ibalizumab (TMB-355), which it is manufacturing for Taiwan Taimed Biologics In

"Taimed is pleased with the speed and excellent execution of this project to ensure the success of this important drug," said Taimed CEO James Chang. "We look forward to t collaboration with Wuxi to successfully bring the drug to marketing approval".

Ibalizumab is a humanized monoclonal antibody (MAb) that belongs to an emerging class of HIV therapies called viral-entry inhibitors. It binds to the CD4 molecule, the primar infection, to interfere with the penetration of the virus into the cells. It is the first entry-blocking humanized MAb used to treat HIV infection.

BioCentury

Home > The Daily Extra > Company News > Chinese manufactured biologic gets approval for U.S. use

Chinese manufactured biologic gets approval for U.S. use

Published on Monday, May 5, 2014

TaiMed Biologics Inc. (GreTai:4147) and manufacturing partner WuXi PharmaTech Inc. (NYSE:WX) said FDA approved a batch of an IV formulation of TaiMed's ibalizumab (TMB-355) for use under an IND -- the first reported biologic manufactured in China to receive clearance from the agency for U.S. use, according to the partners. FDA said under federal law it cannot confirm if this is the first biologic manufactured in China to receive IND clearance.

nature medicine

NATURE MEDICINE | NEWS

First Chinese-made biologics slated for human trials in the West

Boer Deng

Nature Medicine **20**, 562 (2014) | doi 10.1038/nm0614-562

Published online 05 June 2014

More than 70% of novel biologics are developed in the US or Europe, and they are mostly

ChinaBio® Daily Update

Contents

- WuXi Approved to Supply TaiMed Biologic Drug for US Trials
- More Recent Articles

WuXi Approved to Supply TaiMed Biologic Drug for US Trials

DRUG DISCOVERY & DEVELOPMENT

News

FDA OKs Ibalizumab, a Sterile Bi

Mon, 05/05/2014 - 11:04am

Get today's drug discovery & development headlines and

WuXi PharmaTech and TaiMed Biologics announced that the FDA has approved the first batch of the ibalizumab (TMB-355) drug substance and sterile drug product manufactured

凤凰财经

药明康德为中裕新药生产的Ibalizumab获FDA批准用于新药临床试验

Confidential Information

WuXi Congratulates TaiMed Biologics on Receiving Breakthrough Therapy Designation from U.S. FDA

SHANGHAI, Feb. 27, 2015 -- WuXi PharmaTech (Cayman) Inc. (NYSE: WX), a leading open-access R&D capability and technology platform company serving the pharmaceutical, biotechnology, and medical device industries with operations in China and the United States, congratulates our partner TaiMed Biologics for receiving breakthrough therapy designation from the U.S. FDA for ibalizumab (TMB355). The FDA designates a drug as a breakthrough therapy if it is intended, alone or in combination with one or more other drugs, to treat a serious or life-threatening disease or condition and preliminary clinical evidence indicates that the drug may demonstrate substantial improvement over existing therapies on one or more clinically significant endpoints, such as substantial treatment effects observed early in clinical development.

Ibalizumab is a humanized monoclonal antibody and a member of an emerging class of HIV therapies known as viral-entry inhibitors. It is being developed by TaiMed Biologics for the treatment of HIV/AIDS infection. Clinical-trial supplies of ibalizumab were manufactured at WuXi's biologics manufacturing facilities. WuXi will work closely with TaiMed in additional clinical-supply manufacture and process validation to assist TaiMed's anticipated submission of a rolling BLA application. If approved, ibalizumab would be the first biologic product manufactured in China to be launched in the U.S. market.

"We commend our partner TaiMed for this significant regulatory action," said Dr. Ge Li, Chairman and CEO of WuXi PharmaTech. "This is the fourth WuXi-manufactured product to be so designated, reflecting the high quality of the molecules that we choose to work on. We will give TaiMed our full support as they advance this promising product candidate through regulatory review."

Summary II

- Process successfully transferred to WuXi Biologics and optimized for SUBs
- Process scale-up & confirmation in 50L, 250L SUBs
- Tech transfer & scale-up to 2,000L SUB for GMP production
- First biologics produced in China for clinical trials in the US
- Disposable bioreactor/technology allows rapid deployment at GMP facility
  Lack of scale-down model could be a challenge
  Close collaborations with vendors required
- Comparability protocol governed the whole process transfer, optimization and manufacturing

Technology Transfer to US FDA approval for clinical use: < 20 months (Sept 2012-April 2014)

Topics 4: One-stop ADC Services: From Ideas to Manufacturing

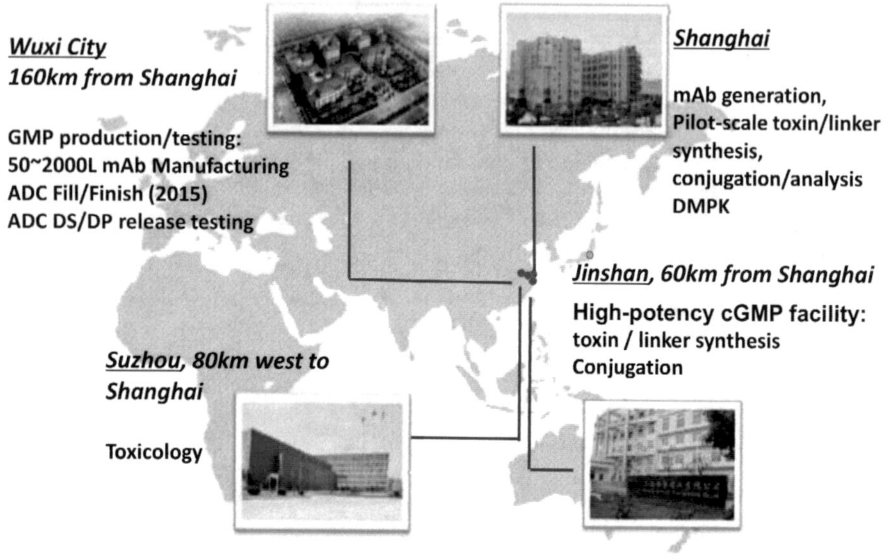

Fully Integrated Service for Global Development of a Novel ADC

Ambrx and Zhejiang Medicine Co. Ltd. Form Collaboration to Develop and Commercialize Ambrx's Antibody Drug Conjugate for Breast Cancer

PRNewswire **Press Release:** WuXi PharmaTech (Cayman) Inc. – Sun, Jun 16, 2013 11:05 PM EDT

Email | Recommend 5 | Tweet | Share | +1 | Print

RELATED QUOTES

Symbol	Price	Change
WX	20.13	+0.62

SAN DIEGO and SHANGHAI, June 16, 2013 /PRNewswire/ -- Ambrx and Zhejiang Medicine Co. Ltd. (ZMC) today announced that they have formed a collaboration to develop and commercialize ARX788, Ambrx's most advanced internally developed site-specific antibody drug conjugate (ADC) targeting Her2-positive breast cancer.

Under the agreement, Ambrx and ZMC will continue the development of ARX788, with ZMC bearing the ongoing development cost. ZMC will receive commercial rights in China while Ambrx retains commercial rights outside of China and receives royalties on sales of the product in China. ZMC will manufacture the product to world-class standards for clinical and commercial supplies on a global basis. WuXi PharmaTech will provide integrated services for ARX788, including the development and manufacturing of the toxin, antibody and ADC, pre-clinical development and clinical trials.

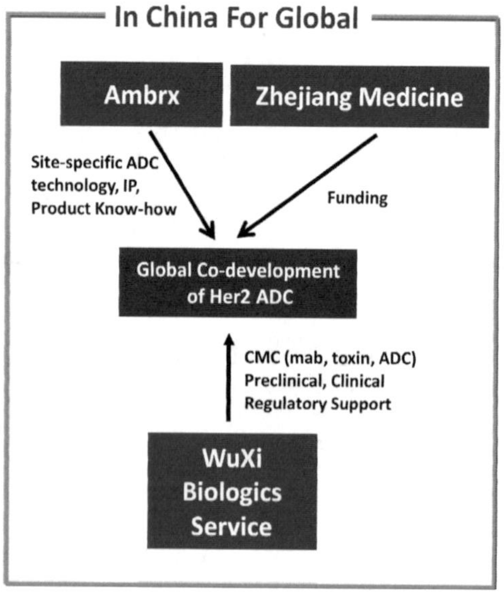

Concluding Remarks

- **WuXi Biologics has developed a leading open-access integrated biologics development and manufacturing platform to expedite globe development**

  Diversified technology platforms enable achievement of a balance among quality, speed and cost for different product modalities: mabs, fusion proteins, recombinant proteins, ADCs

  Use of disposable technologies for large-scale manufacturing provide flexibilities: Scale up by numbers of 2000L bioreactors for commercial manufacturing

- **Integrated services from ideas to manufacturing enable speedy development and manufacturing by avoiding unnecessary transfers and reducing timeline risks**

Acknowledgements

- WuXi Biologics Process Development Teams
- WuXi Biologics MFG/QC/QA
- WuXi Biologics Project Team
- WuXi Biologics Management Team
- Vendors
- Clients

SHANGHAI HONGQIAO RAILWAY STATION

Small scale installations meet 10's to 1.000's kg demand for modern biotherapeutics

AccBio 2015
Scottsdale, March 2015
Günter Jagschies, GE Healthcare, Sweden

Imagination at work.

Amgen Singapore – Facility of the Future

In Nov 2014, Amgen celebrated the inauguration of a ***$200 million plant in Singapore that incorporates continuous processing***. The new 120,000 sqft plant will produce a drug substance used in the treatment of osteoporosis and bone-related cancer. The company plans to employ ***200 people***.

The plant in Tuas uses **single-use bioreactors, disposable plastic containers, continuous purification processing and real-time quality analysis for monoclonal antibody manufacturing**. Started in early 2013, the **plant was constructed in just 15 months,** about half the time it would have taken to build a conventional manufacturing plant, and is about ***75 per cent smaller*** than a conventional biologics facility but ***can manufacture a ton a year***.

The processes at the plant are more cost-effective: it ***costs only a 25% of the capital cost and one-third of the operating expenses at conventional plants, and uses 80% less energy and water***.

- 6x 2.000 L single-use bioreactors (XDR, GE Healthcare), same scale in development and manufacturing
- Chromatography columns with intelligent packing: methods validated and stored in column (AxiChrom, GE Healthcare)
- Interconnected steps in downstream processing ("continuous")
- Electronic batch records
- Much reduced off-line QC testing
- No redundancy in facility

Kimball Hall, Amgen at AccBio 2015

The AccBio Triangle

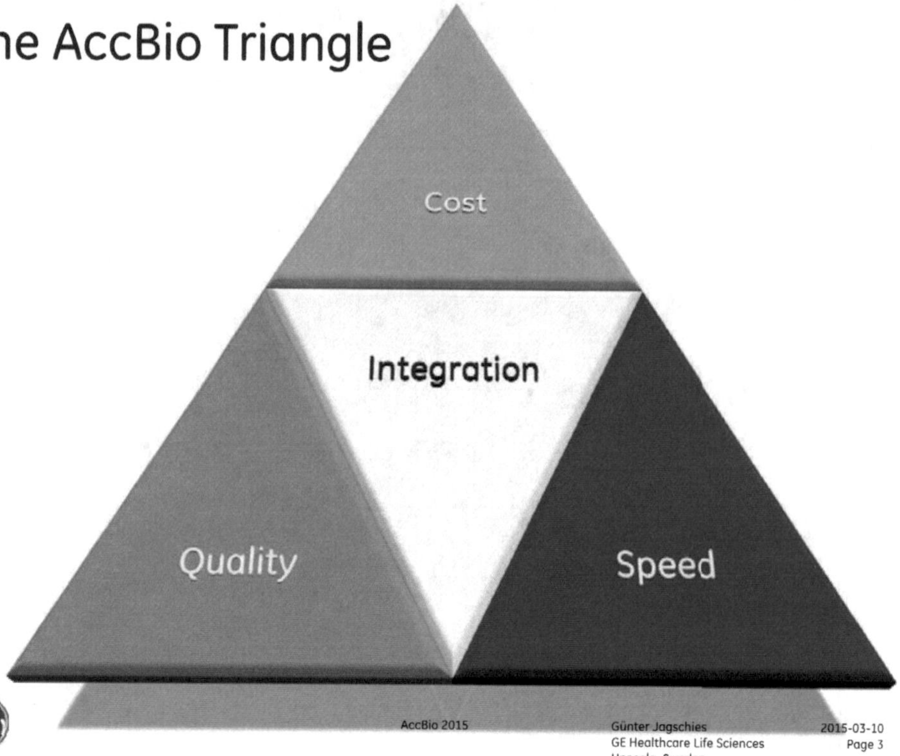

MAbs et al. large scale?

Production rate estimate: product revenue to manufacturer is 50% of pharmacy list prices, 30% of manufactured product cannot be sold due to various inefficiencies in the supply chain

Product	Marketing companies	2012 revenue [$ million]	Worst case# production [annual kg BDS]
Humira	Abbvie	9.522	648
Enbrel	Amgen / Pfizer	7.973	1.453
Rituxan / MabThera	Genentech / Roche	7.153	2.047
Remicade	Janssen / Merck	7.148	1.239
Herceptin	Genentech / Roche	6.280	1.613
Avastin	Genentech / Roche	6.147	1.894
Synagis	Astra Zeneca	1.880	146
Erbitux	BMS / Imclone / Merck Serono	1.842	674
Tysabri	Biogen Idec	1.631	214
Xolair	Genentech / Novartis	1.256	449
Prolia / Xgeva	Amgen	1.220	172
Orencia	BMS	1.176	909
Stelara	Janssen / Merck	1.025	13
Simponi	Janssen / Merck	938	64
Actemra	Genentech/Roche	898	453
Yervoy	BMS	706	10
Vectibix	Amgen	610	136

There are many products and scenarios where the legacy scale seems inappropriate

Originator products
- Six ton-scale products
- More potent APIs have lower scale at same or higher revenue
- Most newer products at 10s-100s kg scale

Five biosimilars of a 2-ton original: at global 20% biosimilar market share: high 10s to low 100s kg scale per winning biosimilar brand...

...lower initially and much lower for local products.

Rec Proteins – a great small volume story

(Production volume and revenue map, 2006 data)

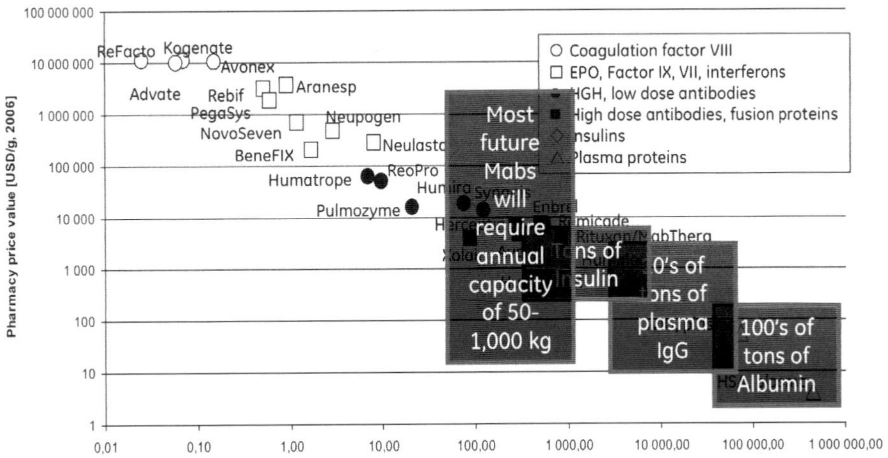

- Pharmacy list prices 2006, available packages: Arzneimittelkompendium (CH), MediMedia Gelbe Liste Pharmaindex (GER), RxUSA.com Pharmacy (USA)
- Annual global sales 2006: Annual reports 2006
- Plasma products: Marketing Research Bureau 2004

Günter Jagschies
GE Healthcare Life Sciences
Uppsala, Sweden

5

Market demand

"Just" have a ton-scale blockbuster (or multiple large products)...

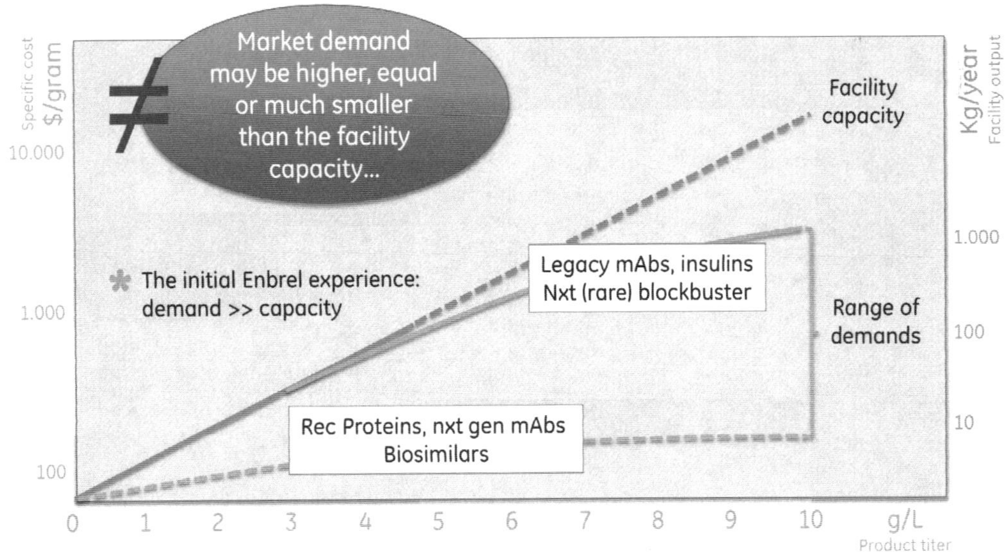

AccBio 2015

Günter Jagschies
GE Healthcare Life Sciences
Uppsala, Sweden

2015-03-10
Page 6

Is it all just about titer?

"Just" implement a 90% cost reduction...

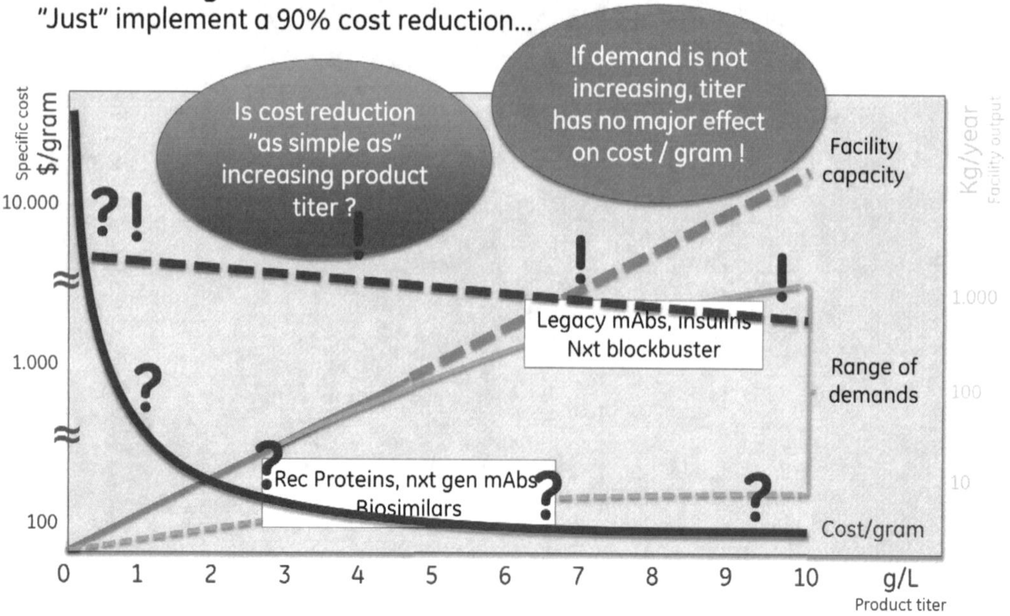

AccBio 2015

Günter Jagschies
GE Healthcare Life Sciences
Uppsala, Sweden

2015-03-10
Page 7

Beyond improvement of "islands"

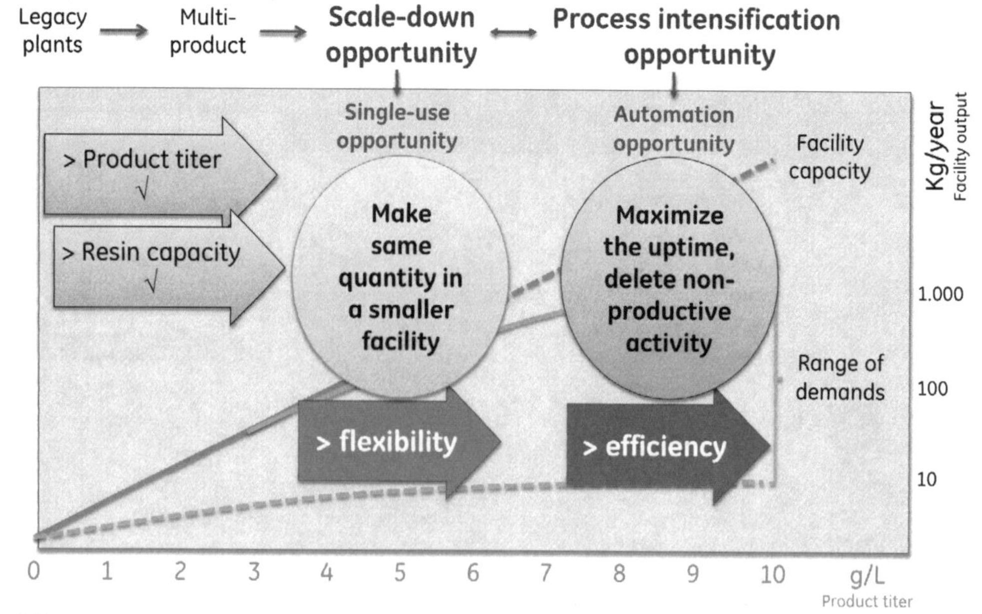

AccBio 2015

Günter Jagschies
GE Healthcare Life Sciences
Uppsala, Sweden

2015-03-10
Page 8

Dimensioning: batch size capability

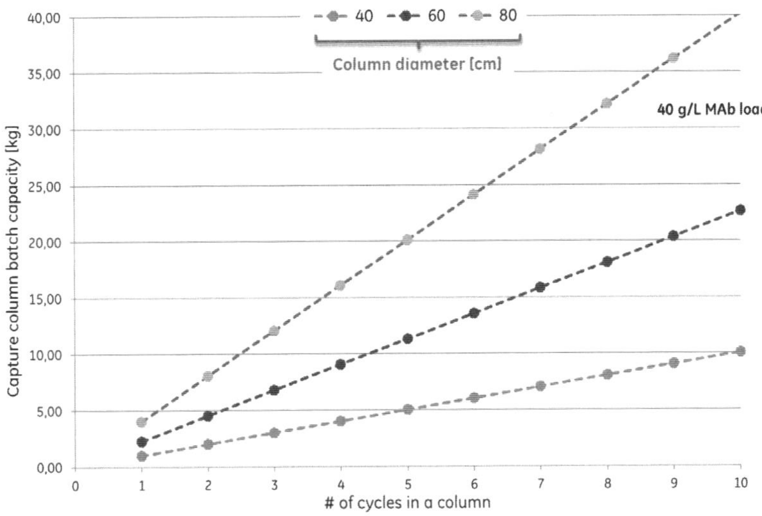

40 g/L MAb loaded on capture column

AccBio 2015

Günter Jagschies
GE Healthcare Life Sciences
Uppsala, Sweden

2015-03-10
Page 9

Dimensioning: batch size capability

Small to mid scale columns easily handle high titers

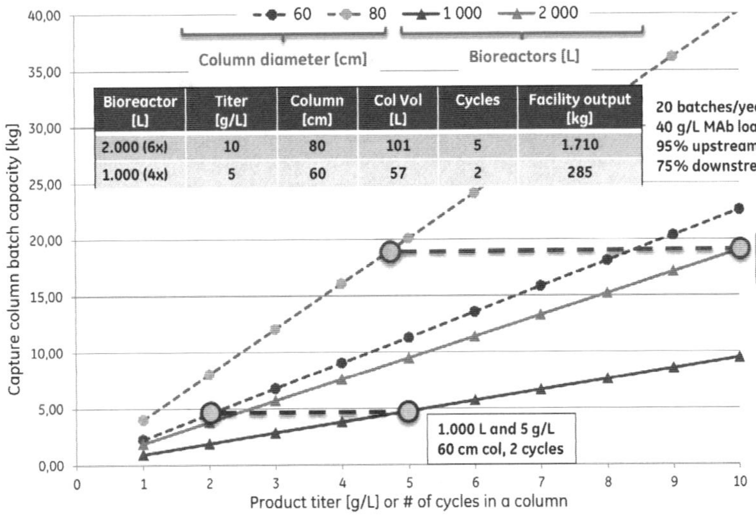

Bioreactor [L]	Titer [g/L]	Column [cm]	Col Vol [L]	Cycles	Facility output [kg]
2.000 (6x)	10	80	101	5	1.710
1.000 (4x)	5	60	57	2	285

20 batches/year/bioreactor
40 g/L MAb loaded on capture column
95% upstream yield
75% downstream yield

AccBio 2015

Günter Jagschies
GE Healthcare Life Sciences
Uppsala, Sweden

2015-03-10
Page 10

3-step purification platform for mAbs

Purification steps with latest generation resins

- Titer: 5.5 g MAb/L
- Aggregate conc.: 2% to 3%

add flocculation step?

MabSelect SuRe LX → Capto S ImpAct → Capto Q (FT)

MabSelect SuRe™ LX
- High load capture on Protein A

Capto™ S ImpAct
- Aggregate removal at high load

Capto Q used in flow-through (FT) mode
- HCP and DNA removal

6x 2.000 L facility capability

Current best case at 10 g/L product titer, full utilization

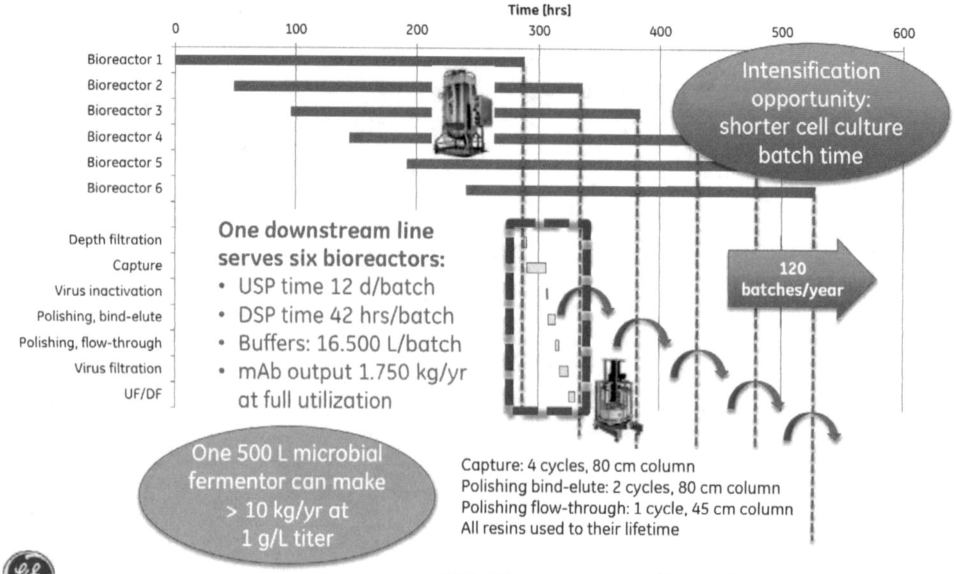

Further process integration options

Improving scheduling flexibility:

- Decouple seed train from bioreaction
- Decouple harvest from purification

Reduce time for non-productive activities:

- Design for connected processing
- Design for continuous processing
- Eliminate up to 70% of buffer prep effort

Simplify the process design:

- Reduce number of buffers (3-5)

Small facility investment cost

Published CAPEX values: large stainless steel plants

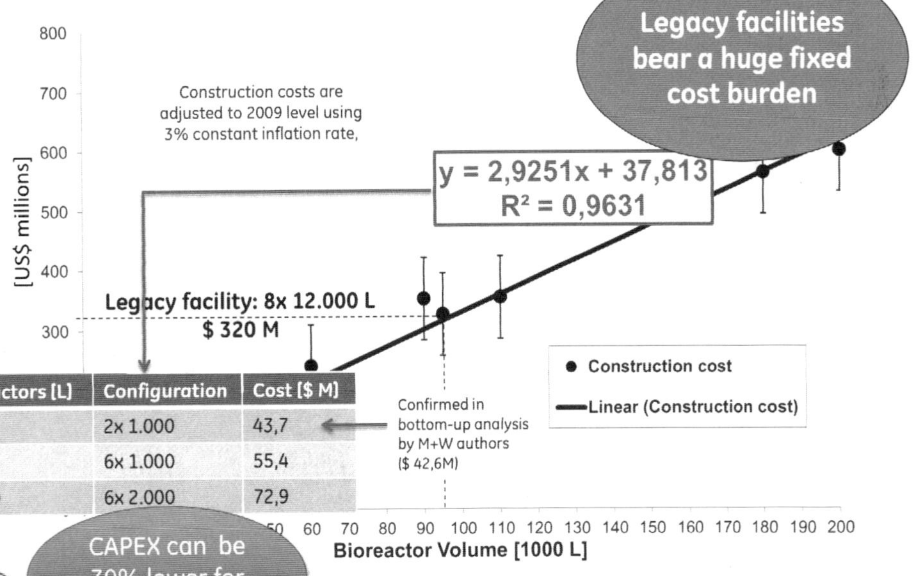

Bioreactors [L]	Configuration	Cost [$ M]
2.000	2x 1.000	43,7
6.000	6x 1.000	55,4
12.000	6x 2.000	72,9

CAPEX can be 30% lower for single-use facility

Objective: $ 100 / gram (mAb model)

"Work in progress" economic analysis

Scenario	Direct capital	Cost target	Min production	Min titer
SU 2x 2.000 L	$ 30M	$ 100/g	185 kg	~3,5 g/L
SU 6x 2.000 L	$ 50M	$ 100/g	300 kg	~2 g/L
SS 6x 2.000 L	$ 73M	$ 100/g	360 kg	~2,5 g/L
SS 6x 15.000 L	$ 300M	$ 100/g	960 kg	~1 g/L

- Investment costs from previous graph
- Minimum annual production rate to get to cost target assuming variable cost of $ 25-35/g
- Minimum product titer to get to target production cost assuming full utilization (max batch number in facility
- Other assumptions:
  - 50% higher labor cost in 6x facility over 2x
  - 15% higher labor cost in 6x 15K as compared to 6x 2K facility
  - Analytical cost is fixed not variable

AccBio 2015 | Günter Jagschies, GE Healthcare Life Sciences, Uppsala, Sweden | 2015-03-10 | Page 15

Conclusion - achieving cost targets

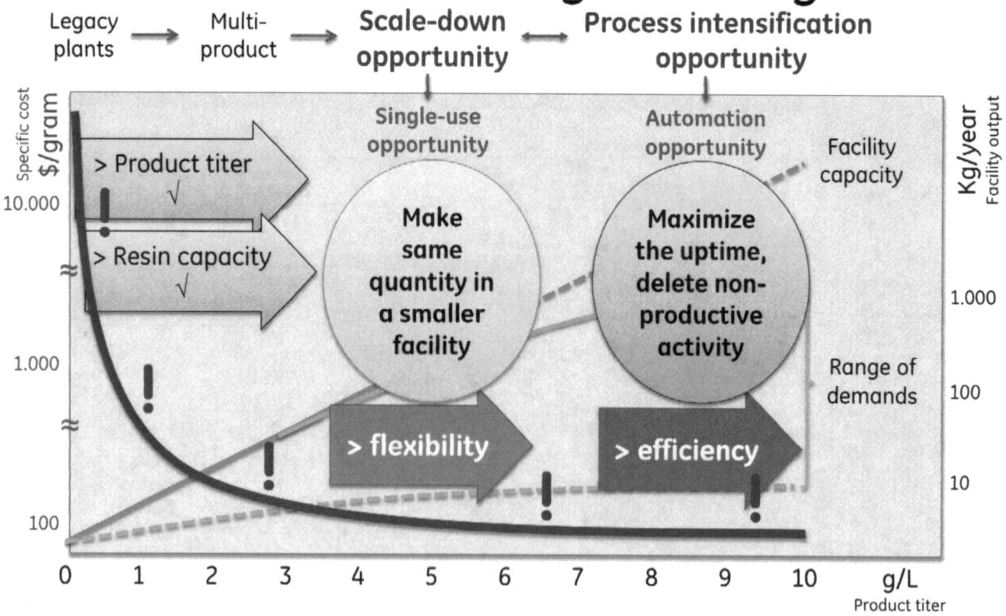

AccBio 2015 | Günter Jagschies, GE Healthcare Life Sciences, Uppsala, Sweden | 2015-03-10 | Page 16

Conclusions

- Small bioreactor installations up to 2.000 L scale cover the market demand for mAbs (10's to 1.000's kg)
- Bottleneck in DSP effectively removed: columns between 40 – 80 cm diameter, intelligent columns remove packing issues
- CAPEX demand is 5x lower than for legacy scale facility
- Single-use technology boosts flexibility and offers additional cash flow advantages
- Classic scale up replaced by "scale-out"
- Investment into capacity (closely) follows market demand rather than precede it by many years

The AccBio Triangle

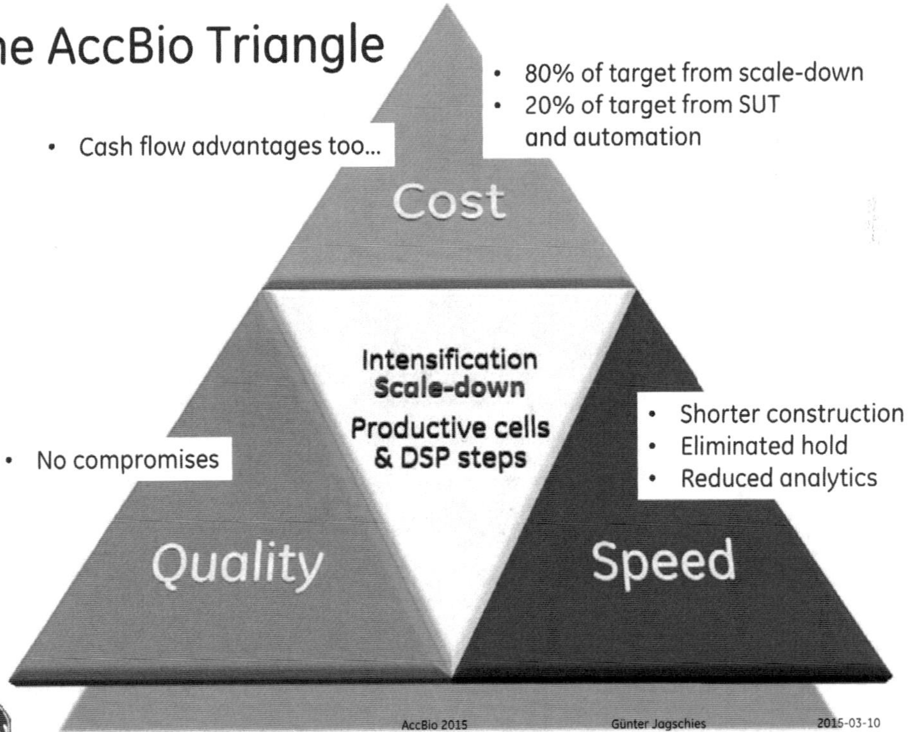

www.gelifesciences.com/Bioprocess

GE Healthcare Bio-Sciences AB
Björkgatan 30
751 84 Uppsala
Sweden

imagination at work

A Process Development Strategy Providing Flexibility While Balancing Cost and Speed without Sacrificing Quality

Allen, M.,

Originally presented on: 3/4/2015 1:40:00 - 2:05:00

Platform process development approaches work well for large molecule therapeutics including monoclonal antibodies (mAbs). Pfizer's large molecule pipeline includes a variety of modalities ranging from mAbs to antibody drug conjugates to fusion proteins and native molecules. To efficiently prosecute this diverse portfolio we apply approaches that balance risk, timeline pressures, and cost without sacrificing quality. To accomplish these objectives we first assess the manufacturing risks associated with each therapeutic candidate molecule by conducting a molecule assessment, a combination of experimental and *in silico* evaluations designed to provide an early understanding of the fit for our Phase 1 process development platform.

For mAbs deemed "well behaved" we can apply a minimalist process development strategy designed to progress the therapeutic candidate to manufacturing for IND enabling toxicology studies as rapidly as possible. In these cases we apply a defined cell culture platform, a well-established 2-column purification process and platform analytical and formulation approaches. Initiation of cell line development activities well in advance of final candidate identification further reduces the timeline from candidate selection to manufacturing. This is accomplished by employing a targeted integration expression strategy. The targeted integration strategy allows parallel development of multiple molecules in a resource sparing manner, but may suffer from less than optimal expression requiring post-Phase 1 process optimization. To address this liability, we aspire to further increase efficiency by establishing a next-generation targeted integration host cell line that we expect will increase expression and reduce resource utilization.

For molecules deemed "not well behaved" we apply more extensive process development strategies with extended timelines and increased costs. In all cases, however, we apply lessons learned from previous experience to reduce or eliminate potential issues, including amino acid sequence liabilities and DNA sequences known to be prone to mis-splicing. Additionally, we apply robust genetic and biochemical sequence variant analysis to ensure the amino acid sequence fidelity for each therapeutic candidate. Taken together these strategies enable an early understanding of the most efficient process development approach that balances cost, speed, and quality of the product and the process.

AUTHOR INDEX

AIChE
120 Wall Street, FL 23
New York, NY 10005-4020

ISBN 978-1-5108-1578-0

CURRAN ASSOCIATES INC.
proceedings.com